chronos

ÉTIENNE KLEIN
Translated by Glenn Burney

chronos

HOW TIME SHAPES OUR UNIVERSE

THUNDER'S MOUTH PRESS
New York

CHRONOS
How Time Shapes Our Universe

Published by
Thunder's Mouth Press
An Imprint of Avalon Publishing Group Inc.
245 West 17th Street, 11th Floor
New York, NY 10011

Originally published as *Les Tactiques de Cronos,* Éditions Flammarion, 2003
First Thunder's Mouth Press edition November 2005

Library of Congress Cataloging-in-Publication Data is available.

ISBN: 1-56025-708-3
ISBN 13: 978-1-56025-708-0

9 8 7 6 5 4 3 2 1

Book design by by India Amos
Printed in the United States of America
Distributed by Publishers Group West

To Paul and Jules, whose youthful laughter defies time

contents

introduction

Historians of science agree on one point: "modern" physics truly begins with Galileo's discovery of the law of gravity. But until very recently, it was impossible to celebrate this important anniversary, because no one knew the exact date of Galileo's discovery. The law of gravity did not appear in his 1590 treatise *De Motu* ("On Motion") but was clearly explained in his famous *Dialogue Concerning the Two Chief Systems of the World,* published in 1632. And while Galileo left a mountain of scientific notes as he prepared these two books, all of these notes are undated.

In the spring of 2002, however, physicists at the National Nuclear Physics Institute in Florence used an ingenious method to determine the approximate date Galileo set down his famous law. By projecting a beam of protons at his notes, they were able to measure the quantity of iron, copper, zinc, and lead in the ink.[1] They then determined that the ink Galileo used when he wrote the first law of gravity came from the same batch of ink used to write his accounting notes dated 1604. Physics thus helped to write its own history with enough precision for us to finally celebrate its birthday. In 2004 "modern" physics reached the four-century point.

Galileo's discovery of the law of gravity opened the doors of physics to the concept of time and radically changed the way humans depicted it. Until then, time had been strictly considered along with everyday events. It essentially served to help men orient themselves in the social universe and regulate their (co)existence with the global flow of earthly events. When Galileo managed to determine the status to be given to time in order to make measuring movement possible, he found a true science of dynamics. By studying falling bodies, he eventually determined that if time rather than distance traveled were chosen as the variable, then a body falling through space obeyed a very simple law: the speed attained is proportional to the duration of the fall and independent of the mass and nature of the falling body; thus, a kilogram of lead drops exactly like a ton of iron. The major upshot of this discovery was to contradict Aristotle's theory of movement, which for two interminable millennia postulated that the speed of a fall increased with the body's mass. In addition, Galileo's discovery established the first mathematical expression of time, upon which Newton would later found his laws of mechanics. The new law of gravity thus brought down the old ruling corpus.[2]

This story has a moral: certain scientific discoveries have enough impact to undo entire chapters of a dominant philosophical system. It also demonstrates that modern physics and time are linked. This is not to say that physics, perched on some Mount of Knowledge, is *the* science of time, nor that it enjoys some privilege allowing it to impose its own logic as it sees fit, but there is obviously a clear affinity between the two, a fertile complicity. From the moment Galileo tamed time by making it a mathematic variable, physics has discussed and theorized about time in ways we could never have guessed without its help.

Four centuries have gone by since Galileo's day, during which the earth has turned around the sun many times. Physics has benefited by taking the time to grow, consolidate its position, and then trigger revolutions at a breathtaking pace, especially during the twentieth century: Einstein's special theory of relativity, quantum physics, general relativity, the discovery of nuclear forces and antimatter, the expanding universe, and others—so many landslides have occurred in the field of basic concepts.

Each of these radical changes has in its own way challenged time's previous status. Along the way, physical time has lost some of its imagined purity and much of its independence; it now finds itself inseparably linked to space, associated with energy, and anchored in matter. Actually, we can hardly recognize good old time, which is now implicated in improbable liaisons that transform its identity. But none of these liaisons has remained without offspring: whenever it had to deepen its conception of time, physics improved its operating efficiency, took over uncharted territory, discovered new phenomena. It's as if each progressive step in the theorization of time translated into immediate dividends. In order to resolve a time-related problem, theorists in the 1930s were led to predict the existence of antimatter! Did time really become the "main matter" of physics? Will physics one day manage to grasp its very substance? It's still too early to say.

Yet one thing is certain. When it comes to time, contemporary physics destroys common ground, shakes the vulgates, and expands the horizon. Boosted by recent successes in particle physics and cosmology, it does not hesitate to "play" with time, to formulate audacious hypotheses (the discontinuity or plurality of time, for example) that would appear crazy if strong theoretical arguments did not allow us to picture them. In this way, physics revives indestructible questions, sheds light on them, and at the same time creates new questions. Did time appear "at the same time" as the universe, or did it precede it? How did it start? Who initially flicked the switch? Is it of the world, or does it contain it? What does time consist of, this time that passes but is always there, that doesn't change but changes all things? What is its real relationship to things? Does it exist independently of all that appears, changes, wears down, ages, and dies? Are these famous "wormholes" really time-traveling machines? In what ways does the superstring theory turn our conception of space and time upside down? What are the points of convergence between physical time and lived time?

Some of these questions, until now left to the realm of metaphysics, today find themselves in the realm of physics. This radical shift seems to provide arguments to those who believe a new Einstein could soon gain a complete and definitive understanding of time. Such an illusion or error does not come out of the blue. The conquests of contemporary physics are so notable that they fuel the hope of one day being able to put an end to some of the "big" questions out there, including the nature of time.

Until then, the mathematical time of physicists—in their eyes, the only "real" time—has very little in common with our usual sense of time. It is but a small step from there to think that time spends its time using its cunning to abuse us, or that we spend our time confusing it with our perceptions of temporal phenomena. We mask time with fallacious properties, permanently hiding it. Therefore, the main objective of this book will be to expose the tactics of the *tick-tock* of our watches through which time hides its true nature. By showing its face, time actually hides, because this actor principally uses stand-ins.

Actually, time always runs the risk of being identified by the phenomena it contains. Yet what happens *in* time isn't the same thing as time itself. Phenomena, as they unfold, cloak time in their own attributes: change, evolution, movement, repetition, succession, death. Time should not be confused with the various displays it makes possible. Time is, as Balzac's Minette said, "a tall, thin thing."[3]

We will thus devote our efforts to various exercises of deconstruction, a sort of thinning out of the leaves of time. At first, our approach will be apophatic, as the philosophers say; we will try to circumscribe time by saying what it is not. How? First, by looking into our often worn-out language. Almost a full century after Einstein's work, we still speak of time in the same way people did before Galileo! It's as if modern physics never existed. Second, we will flush out the paradoxes hidden under the carpet of habits by ridding time of all temporal phenomena that give it substance and end up settling in our way of speaking about it. Finally, we will examine physics in an attempt to isolate time's intrinsic qualities, things that make time be time and that cannot be taken away without throwing the baby out with the bathwater. Things that, through a mirror effect, will teach us how physicists work: are they more with Heraclitus and evolution, or with Parmenides and unchanging being?

When all is said and done, time might not look at all the same as we once viewed it.

does a clock speak for itself?

Have you not done tormenting me with
your accursed time?
It's abominable! When! When!
SAMUEL BECKETT, *WAITING FOR GODOT*

I call on all those lovers of exact time.
EDGAR ALLAN POE

Time for us is a sort of familiar evidence, an obvious being, a reality that goes without saying. We always take it to be there around us, secretive, silent, but constantly at work—in this falling leaf, that child being born, this crumbling wall, that birthday candle we blow out, this blooming love affair, that fading one. Don't these phenomena, all very real, constitute enough tangible manifestations of time? Common experience seems to suffice, so that we can never doubt the existence of time. Isn't time something that simply shows itself on the faces of our watches?

Look at the second hand of a watch as it moves constantly forward. Isn't it showing you time as it truly is—nearly naked, almost pure—through the means of the circular procession of hours, minutes, and seconds? Or in Martin Heidegger's (always somewhat complicated) words, is it in "the presentification of the advancing arm" that time makes itself most clear?[1] A watch is an object that by definition and purpose shows something other than itself. What does a watch show? Time, of course! That's the unequivocal answer from those with quick tongues. However, if we think it over a bit further, this supposed demonstration of time through clocks of all sorts isn't a given.

But what, in fact, do watches and clocks—those oh-so-familiar objects that are such a part of daily life, whose hands symbolize in our eyes time in movement—*really* show? They make the movement of the hands visible, that much is certain. But this regular movement, which admittedly assumes that time is passing, even taking place in space, too hastily identifies itself with time—as if it were explained right there in the *tick-tock* marking its flow. We won't remain duped by this tactic any longer.

We all can agree that a clock gives the hour. It passes its time doing only that. But it doesn't show anything of what time is prior to this process of actualization. A clock more accurately dissimulates time behind this convincing mask of perfectly regular mobility. By disguising it with movement, it shifts it; time becomes a manifestation of space, a stand-in for expanse. But does movement become confused with time? More accurately, movement is a camouflage for time, an ersatz version, though easy to identify; when a clock stops, its immobilized hands do not keep time from flowing. The pause in movement is not equivalent to a pause in time; a motionless object is just as temporal as an object in movement.

Sure, we say, but every timepiece is also a chronometer; it allows us to measure duration. And therefore time? Let's see. Let's first note that each duration is a very strange, even mysterious, "object." Unlike length, its spatial equivalent, duration is never present in extenso, since it consists of instants that do not coexist. It is a quantity that is never truly there, never laid out before our eyes. We can pass through it, live it, and we can measure it by using a watch or clock, but there is not a single duration that can truly be shown or perceived in and of itself.

So does measuring duration equal measuring time? It would be better to say that time is what allows us to have duration. It creates continuity among a series of instants. But the measurement of duration in no way makes time, which invented it, appear. It reveals nothing of the mysterious mechanism that allows one instant to appear only to immediately give way to another instant, ad infinitum.

Time is none other than a mechanism, a machine, that permanently produces new moments. It is an intimate engine, a breath hidden close to the world, through which the future becomes first the present, then the past. It's a secret force by which tomorrow "slips" into today, by setting an exact amount of time for this daily-repeated operation. All durations happen only because of time's unending pushing effect.

In short, all clocks disguise time in a mix of movement and duration, duping us to confuse time with this mix.

Are these "simulations" of time? No, they are more like dissimulations—a strange hand that advances while showing us nothing of what it symbolizes. *Time resides outside of timepieces.*[2] More precisely, there is scarcely more time to see inside than outside a watch, for the simple reason that time in no way exposes itself directly: we have never seen, felt, heard, or touched a time that produces the succession of instants. It never appears as a raw phenomenon. We only perceive its effects, its deeds, its attire, its manifestations, things that can fool us about its nature.

History teaches us that the measurement of duration precedes any elaboration of the concept of physical time. Several millennia separate the invention of the first clocks, which were quite laborious, and Newtonian time. All too often we imagine that time was only measured by means of inanimate instruments like sundials, hourglasses, clepsydras, or mechanical clocks. And yet, people also used animals for the same purpose. So on one wall in Tutankhamen's tomb twenty-four baboons represent the cycle of the hours. The ancient Egyptians had indeed noticed that this animal had the peculiarity of urinating at rather regular intervals, nearly every hour. So they used its bladder for a pendulum.

But the "fold of time" really took root in the monasteries of the Western world. In contrast with the variations of secular life, monastic orders set down iron discipline, creating a rhythm that left little time for whimsical pleasure. Well before the first mechanical clocks appeared in the thirteenth century, a papal bull from seventh-century Pope Sabinian, decreed that monastic bells—which generally used a clepsydra, or water clock, and a hammer—had to sound seven times every twenty-four hours. These regular chimes punctuated the day and constituted the canonic hours, marking periods consecrated to devotion:[3] *matins* (night prayers), *lauds* (one hour before sunrise), *prime* (first hour of daylight), *terce* (midmorning), *sext* (noon), *none* (midafternoon),[4] *vespers* (sunset), and *compline* (one hour after sunset).

Little by little, this temporal discipline spread from monasteries to towns. At the beginning of the fourteenth century, clock towers, with their enormous iron and copper constructions, rang out the hours in towns all over Europe, synchronizing human and social activities and thus bringing a previously unknown regularity to the lives of craftsmen and merchants. But during the two long centuries that followed, the passage of time, though duly and precisely measured, did not intervene in any quantitative fashion in the study of natural phenomenon. The appearance of clocks did not, in and of itself, allow physical time to arise in the minds of men.[5]

Though time underlies all things, it does not reveal a glimpse of itself in anything. It remains buried beneath all surfaces. And that is its genius—invisible even to X-rays, it never deigns to surrender itself like an empirical object. However, language unrelentingly invokes it as a familiar object, while no one has seen it face-to-face and it has never signaled its presence.

the word "time"; or, every dictionary's embarrassment

That which we can't speak about,
is that which we must say.

VALÈRE NOVARINA

Get rid of heroism, and it comes back on a bike.

LOUIS NUCÉRA

WHEN WE SPEAK of time, everyone tacitly understands what we are talking about. Who doesn't claim to intimately know what it is? No need to be a Kant, Einstein, or Heidegger to claim one's authority or put forth one's own general idea of the thing. No, siree! We belong to the human condition, we have experience, and that's all it takes, or so we believe, to raise the question of time. So we drag out the old truisms, we recycle fossilized ideas, we elevate simple water-cooler rumors to the realm of collective belief.

The truth is that the notion of time—carried by common usage, cradled by culture, and blunted by habit—seems like a simple thing, at least when seen from a distance. Even though philosophers have always presented it as a tremendous challenge of the intellect—equivalent to the physical challenge of a cyclist taking on the Alps in a sprint—we let ourselves be misled by its hilly aspect. Confident in the pertinence of our ordinary experience, we think we can tackle it on a rusty old bike, without any special performance-enhancing injections or prior training, carrying only a thermos of herbal tea.

Then, one fine day, on a detour of more refined thought or out-of-control daydreaming, everything collapses. Our ideas tumble like a house of cards, and we stand there gaping in shock. Leaning hard on the pedals, looking a little silly, we realize that we do not understand the slightest thing about time, that it is a veritable Annapurna, that its familiarity comes from habit and not elucidation. And in a flash, we discover that it remains impenetrable, mysterious, and fundamental.

Stricken with vertigo, we grab onto the first handhold in sight: vocabulary. Isn't time first and foremost, a word? Yes, but its meaning does not seem to be carved in granite. Is it a synonym for *simultaneity,* as in the expression "He's always doing two things at the same time"? Does it invoke the idea of succession, as in "The time will come when this book will be over"? Or does it conjure up the notion of duration, as in "The writer didn't quite have enough time to finish his book"?

In fact, the word *time* vaguely covers three distinct concepts—simultaneity, succession, and duration—allowing us to talk about change, evolution, repetition, the future, decay, aging, and maybe even death, all at the same time.

Blaise Pascal described the word *time* as being a "primitive" word, in the sense that it belonged to that group of words that are so fundamental that it would be impossible (and pointless) to define them.[1] We could object that there are numerous definitions of time, of which he surely knew: "Time is a moving image of eternity" (Plato), or "Time is the number of motion in respect to before and after" (Aristotle), and more recently, "What happens when nothing is happening" (Giono), or what allows all real things to exist. But—and this is what makes Pascal correct—all of these supposed definitions of time are not definitions; they are merely images, tautologies, displacements, since all of them assume, a priori, the idea of time (giving rise to the old question of logic, how do you *found* a *fundamental, ground* a *grounding*?). To the proliferation of the signified, they respond with a scattering of metaphors. But as Montaigne said, "All we're doing there is substituting one word for another, often more obscure,"[2] so the essential meaning of *time* is thus abandoned in the shadow of language. Believing that we are talking about time, we talk about something else instead, as if we could approach time only by borrowing a more radiant means than itself.

The word *time* is also "primitive" in the sense that for speaking of the world and all that transpires in it, we imperatively need this pilot fish of intelligibility. How could we possibly perceive an object, describe an event, express an emotion, or tell a story without setting them in a temporal framework? Deleting the word *time* from our vocabulary would be like sewing our mouths shut. Just take a quick glance at the place it occupies in literature and philosophy, science and poetry, and most of all in pop songs. Time reminds us that life is short, love is ephemeral, and death is certain, in case we get so distracted by pleasure that we forget. An entire poetics (and certainly metaphysics) of time could be drawn from everyday language as well as from the works of Shakespeare, Dante, Goethe, and Proust. Except that this simple word, which contains such varied experiences, says nothing about the nature of time. In appearance it is as banal as the words *bread* or *table,* but this familiar proximity does not hold up for long.

More than fifteen centuries ago, Saint Augustine had already noted this paradox in a sentence so often cited and commented on that it seems hackneyed: "If I am not asked, I know what time is; but if I am asked, I do not."[3] The result being, as a matter of fact, that if I try to think up a lecture about the nature of time, my argument collapses or frays almost as soon as I begin. If I persist, I have to confront a strange double phenomenon: first, time isn't an object in the usual sense of the term (it does not belong to the same type of reality as a chair); second, language seems to fail when we are faced with the need to speak about time. Yet this necessity nevertheless exists, since time is one of the questions that hound us, that seem to arise unprovoked from our unappeased brains.

To speak about time, we often resort to ready-made phrases that almost always conceal ambiguities. When I say, "I do not have time," I mean to say that I have other obligations; I do not have adequate duration at my disposal to do what is demanded of me. But if I do not have this period I need, isn't it precisely because time, by passing, limits and constrains my schedule? If I do not have time, isn't it because there *is* time? When I say, "time hangs heavy on me," I am signaling that I am going through a difficult or insignificant period, that time seems basically empty to me. Does this mean that the less full time is, the more mass it contains, which is the inverse in the case of mechanical objects? When saying, "I killed time as best I could," I admit that I tried to free myself from a period or moment that fell on me (and felt) like a ton of bricks. Does it mean that time is a hardy, living creature that we would sometimes prefer to see dead? The curious claim of this is that, usually, doesn't *time* kill *us*?

The most stereotypical expression—the one that shatters all records of use—says that time "passes." But isn't this formula already an abuse of language? Who would contest that time is what makes everything happen? Inferring that time itself is passing is to go astray. The succession of three moments of time (future, present, past) does not signify that time follows itself. Moments pass, not time. Isn't it exactly because of time's constant presence that things never cease to pass? Time is indeed paradoxically still: within the temporal passing we can detect the presence of an active principle that remains and does not change. Philosophers have been able to recognize this. Heidegger, for example, stated that "time, in its entire unfurling, is not in motion, and is immobile and at peace."[4] This stillness at work in the very heart of time is clear when we examine the status of the present, a contradictory thing: we say that the present passes, since it is never exactly the same, and also that it does not pass, since we never leave the present instant to join another one. The present is therefore simultaneously ephemeral and persistent; always there but never the same, its advent paradoxically links permanence and change.

In short, by saying that time passes, we confuse the object and its function. Yet what is actually passing is reality, and not time alone, since time never stops being there. We should congratulate Pierre de Ronsard for his clarity on this:

> Time is passing, time is passing, Madame,
> Alas, time is not, but we are passing.[5]

Time is maybe nothing more than the redundancy of unfurling reality. We will spend a great deal of time on this question.

Saying that time passes is also stating, at least implicitly, that "it" exists; it passes, therefore it is. This completely banal expression attributes time with the status of being independent of things and processes. Speaking in this way is to give time an ontological promotion that guarantees the ease with which we generally speak of time, even though its autonomy has never stopped raising a debate among philosophers.

It definitely seems that words do not have direct access to time; they just orbit around it and conceal it. Thus, in order to feel what time is beyond the word that names it, we could draw from the recipe proposed by Ludwig Wittgenstein in his *Philosophical Investigations,* these "language games" that are supposed to free us from the sorcery of certain words by directly showing us what they represent. In one of these games, a master stonemason teaches the trade to an apprentice, trying to convey what a brick is. He describes the material, the shape, the color, but to no avail. Ultimately, he takes a brick in his hand, points to it and says, "This is a brick!"[6] A simple recipe, undoubtedly very effective when it comes to a brick, but if the master had wanted to get his apprentice to understand what time was, what could he have placed in his hand? Nothing concrete, and thus he would certainly experience some frustration.

We can try to convince ourselves that this frustration isn't really frustrating, because a primitive word has no other signification than that which we give it. It thus becomes pointless to ask questions about the truth that the word contains or hides. This strategy doesn't lack arguments. Isn't language filled with a fundamental ambiguity? Doesn't it allow us to say everything, truth and falsehood, good and bad? Far from revealing the truth of the world, words perhaps only refer to other words. Interminably. And thus they levitate over the things they designate without ever touching them, so that we have to stop questioning what lives or lies beneath them and accept that they do not reflect a presupposed truth, that they are just tools to caress the world by talking about it.

But are we capable of resisting the temptation to understand what words hide? In order to dig into the world, we first and foremost have verbs, which provoke our minds as much as they program them. If we continue to struggle over the same questions the Greeks worried about, isn't it because language, practically unchanged through the centuries, always leads us toward them, by some kind of charm or bewitchment? As long as there is the verb *to be* that seems to function as *eat* or *drink* do; as long as there are words like *time, space, thing,* and *emptiness;* as long as there are adjectives like *true, false, real,* and *possible,* we will end up running into the same problems, the same enigmas, again and again and will contemplate what no explanation seems to reach. And so goes thought, whose "magnificent and supreme paradox" (to speak like Kierkegaard) is to keep wanting to discover things that are not always in its grasp.

Even if they are powerless to explicitly speak about the world, even if they do not rely on a "preexperience," words do possess one exclusive virtue: they indicate, herald, and open the way. They are tempting mysteries. Thanks to words, thought redoes, undoes a problem forever. It never gives up. Even when it loses the connection between the signifier and the signified, even when out of juice, thought always finds a good metaphor to bail it out.

For time, it was the metaphor of the river, carried by nature's spontaneous eloquence, that first proposed its services.

an unnaturally flowing river

Each day is a Rubicon into which I yearn to dive.

CIORAN

"TIME IS A river made of events," wrote Marcus Aurelius, emperor and skilled philosopher, unless it was the contrary.[1] This metaphor, implicitly linking time to the idea of flow, has not aged at all. Two millennia later, novelist Frédéric Dard's detective, San Antonio, calls his watch a "dripper," and we become smitten with the idea that time is a kind of liquid that "flows" of its own free will. So smitten, in fact, that we never question this idea. (A peculiarity of doxology is to create an a priori truth and transform an offensive action into a report.) By signifying the present as ever changing, like flowing water, isn't this idea stating an obvious fact?

A river is never the same, since it is composed of continuously renewing elements. This statement applies to us as well. We too are "drowning"; each new second propels us into a new world and a new self. It is therefore possible to put forward the idea (along with Heraclitus) that the single thing that does not change about things and beings is their property of *not changing,* so that nothing can remain identical to itself. From this point of view, change and impermanence paradoxically express a timeless law; always at work, they are a manifestation of eternity. And they give rise to the question: what is this underlying or immanent order that governs a perpetually changing reality? Yet, very surprisingly, our way of speaking about time as a river expresses just the opposite; it links it with lability and mobility. It's time that passes by, we say, not the world, and not us either.

This way of speaking is not neutral, since it hypostatizes the idea of time by implicitly giving it some properties that it does not necessarily have. We should then flush out some of these clandestine a priori properties that are swept along with the river's flow.

First, if time were a river, what would be its "bed"? In relation to *what* would it flow by? What would its "banks" be? We see that the idea of flow surreptitiously postulates the existence of some timeless reality in which time would flow. Time is then oddly "moored" to its opposite, as if clothed in "non-time."

In the case of a river, we know what drives the flow: gravity. Upstream being higher than downstream, water always flows the same way, from high to low. But what makes time flow? No gravity comes into play here. Yesterday, today, and tomorrow are equivalent moments of time in that they are at the same "altitude." Time does not start flowing by falling. But then what pushes the present to flow toward the future (that is, unless the future is approaching us)?

Finally, to say that time flows like a river implies that it has a certain speed relative to its hypothetical banks. In everyday language, it is constantly granted the property of speed. Don't we say that time is always "going by faster and faster"? But speed is, generally speaking, the derivative of a certain quantity in relationship to time. The speed of time is then obtained by determining the rate of the variation of time in relation to itself. Barely under way, we have already run into a wall. We will have to get used to the fact that time takes malicious pleasure in turning the simplest terms into terrible traps.

Curiously, though, these paradoxes have not prevented the river metaphor from accompanying the history of thinking about time. Today, we still speak about time like swimmers carried away on a river, the same river that Heraclitus already noticed twenty-five centuries ago; we never step into it at the same point twice.

This is proof that some things do repeat identically, but only by word of mouth.

4

the time before chronos

Myth is a word chosen by history;
it doesn't shoot up from the "essence" of things.

ROLAND BARTHES

Small fish will become big,
provided God grants them life.

LA FONTAINE

"In the beginning," as the most ancient myths say, was an original world that existed outside the bonds of time. Time arrived onstage only after "a certain time" to initiate a genesis, begin a process, provoke an evolution. In these stories, where marvels triumph, time's primary function is not to make the world persist; rather, it is identified with becoming, not with what holds the world within a continuous present. Only by confusing time and becoming is it possible to imagine a stagnant, prechronic world before time, which only comes along afterward to initiate the plot of history.

This merry mix is still alive in our culture. It implies that no time is present if any change is happening; only the future, not duration, would need time.

Let's have a look at the Greek myths.[1] In the beginning were the sky, Uranus, and the earth, Gaia. The sky was born from the earth and then covered her entirely. He gave the earth no breathing room, keeping her in perpetual darkness and pouring himself into her womb. To speak clearly, his only activity was sexual, and Gaia was pregnant with many children, including the Titans, who were living exactly where Uranus had conceived them. There was no space between Uranus and Gaia, which would have enabled their children to come to light and lead autonomous lives.

But Gaia could no longer stand carrying her children in her womb. Unable to come out, they caused her to expand and nearly suffocate. Kronos, the last conceived, then agreed to help his mother and confront his father. While Uranus was lying with Gaia, Kronos grabbed his father's genitals firmly and chopped them off with a sickle made by his mother. Uranus howled in pain (a high-pitched howl, we can easily imagine), abruptly withdrew himself from Gaia, and flew away. Then he settled on top of the world and never moved again. By castrating Uranus, Kronos took a major step in the birth of the universe: he split the earth from the sky and created an open space between them. From that point on, anything the earth produced would have space to develop, and everything living beings gave birth to would be able to breathe, live, and procreate.

That's how future time appeared—blossomed—right after space. As long as Uranus was weighing down on Gaia, generations were not possible, since they remained buried inside the being that produced them. Actually, contrary to what the myth tells us, time already existed, since Uranus and Gaia "suffered" the feeling of duration. But this was a kind of time locked within itself, a kind of time that did not enable anything other than the world's stagnation. Thanks to Uranus's withdrawal, the Titans were suddenly able to leave their mother's womb and began giving birth too, starting the succession of generations. Kronos's emancipation released Chronos.[2] Transporter of the future, opener of history, he eventually could spread infinitely. (Timaeus's myth, in which Plato sets out his cosmogony, also depicts a demiurge who "sets up" time so that the future can take part in eternity.)

Now if you look at Hinduism, you will find similar stories in very ancient books; there too, time only appeared after a certain while, when a particular character arrived onstage. In other words, first there was a world in which, we are told, there was duration but no time! We read in the *Brahmanas*, "In the beginning, only the Waters (the Oceans) existed. The Waters wondered, 'How are we going to procreate?' They made an effort. They tried with fervor and within Them an egg developed. Time of course did not exist then, but the egg floated for the length of a year. During this year, a being appeared, Prajapati."[3] Prajapati, "the lord of creatures," created earth, space, and sky by uttering one- or two-syllable baby sounds. The contradiction between the fact that time "does not exist" and that the egg floats "for a year" does not seem to break the rules of the story, since time is not initially thought of as what produces succcession or duration.

In the East as well as in the West, we were able to tell stories with chronological markers and concurrently assert that time itself did not need "markers." This narrative freedom, which couldn't care less for coherence, tends to forget that time affects beings in their immobility as well as in their future. That is, a being *also* acts when nothing happens: *the future presupposes time, but time does not imply the future.* Thus, it might be that a certain kind of literature confines or restricts time to a very precise function.[4]

Physics divides time from the future, the flow of time from the arrow of time. The flow of time designates the fact that "time passes" and, by passing, creates duration and only duration. In short, the flow of time simply creates the succession of events. As for the arrow of time, it refers to the possibility that things will *become,* undergoing (sometimes irreversible) changes or transformations through time. We will see later that this is not a property of time itself but of temporal phenomena.

To sum up, the flow of time guarantees the continuity of the world, while the arrow of time creates indelible stories and events in it. Does this mean that if the flow of time happened to stop, everything would disappear?

5

the stopping of time; or, the abolition of the world

Estragon: I had a dream.
Vladimir: Don't tell me.
Estragon: I dreamt that . . .
Vladimir: DON'T TELL ME. . . .
Estragon: . . . that time stopped.

SAMUEL BECKETT

We often tend to speak about time as if it could sometimes stop, as if it could even stop existing altogether on occasion. Stopping time is a very deep desire, a very human dream: to protect oneself from the flow of time and hold on to the happy moments. Lamartine wished time would stop flowing; thus, a beautiful day would "last forever." Winnie, from Beckett's *Happy Days*, longed for "the exquisite hour which carries us away" to be transformed into an "unburstable bubble." Each of us has dreamed of a timeless world where nothing would degenerate anymore, where the garden of things would blossom, sheltered from historical events, in a kind of blissful eternity.

But how can this exist in practice? Erwin Schrödinger, the famous physicist and a great lover of women, explained that all you need to stop time is a sincere kiss. He once wrote, "Love a girl with all your heart and kiss her on the mouth; then time will stop, and space will cease to exist."[1] This recipe has been praised and spread since the dawn of time: for those who want to free themselves from old Chronos's tyranny, love always seemed to be a promising if not efficient means.

Only science fiction writers are able to compete with this. In some of their books, they include instances where time really stops passing. Their trick? The confusion between time and motion, a little like the ancient myths. Their stories typically go as follows: the hands of all watches suddenly freeze (which is definitely possible), and we immediately draw the conclusion that time has stopped passing (which is more difficult to swallow). The first lines of Jean Bernard's *The Day When Time Stopped* are surprising: "On May 24th 2006, a Friday, at eleven o'clock, twenty-seven minutes and thirty-four seconds, time stops. Raymond, standing on a sidewalk around the big square, was just resetting his watch. The hands stand still. He shakes his watch. The hands stand still. . . . The traffic lights don't change, some remain red, others green. Cars, buses don't run anymore, frozen in place. The cyclist who was pedaling loses his balance and falls."[2] In this situation, the time engine can run out of gas, yet the world keeps on turning! And good Raymond can even shake his watch in it!

Basically, we are requested to simultaneously accept that time is not passing anymore, but that the world keeps on turning (as if nothing happened) and that in this world where time has stopped, movements are still possible. That's a lot to ask. Because for the world to continue existing, time needs to be there to pass and make the world go on. Even if nothing in this world is happening anymore, even if nothing is moving anymore, time remains active to keep *making it be.* It is the artisan of its own perseveration: at each instant, *it is the one who holds "now" by the hand to make it go through the present.* If time really stopped, this would signify not only the immobilization of everything, but also the immediate interruption of the present, meaning the disappearance of everything that exists. Such a shift to nothingness, instantaneous and complete, would make any apocalypse look like a romance novel; a stop in time would be a death sentence for the world itself.

If time is really this "being" that we never directly meet but that contains everything we meet, there cannot be a world without time. Time is consubstantial with the world: nothing can happen or persist outside of it.

To summarize, time is *at least* the how by which things remain present. Without it, everything would happen at once; no sooner having appeared, the world would dive back into nothingness. We can feel as if it does not pass anymore, but this is only an impression, an illusion, a way of speaking; it never stops passing. Time is not a pond.

There is thus a distortion between our feeling, which adapts very well to the idea of time stopping (or asks for it to be put on hold), and our intellect, which cannot manage to envision anything other than a stop in motion. This statement does not of course guarantee that time will never stop, but it is nevertheless a good example of our incapacity to conceive it, or more precisely, to think of what this stop would actually mean—a fall in nothingness. As soon as we picture nothingness, we make "something" out of it, which it cannot, by definition, be. Everything happens as if we were only able to think of one thing's absence through something else's representation. In our mind, abolition first means substitution. The idea of absolute nothingness thus becomes destructive in itself. Undoubtedly, since nothingness is a blind spot in our minds, we are incapable of conceiving what a stop in time would be.

As soon as "something" is there, there has to be some time, even if it seems that there is no dynamic at work in this "something." In a static universe, of ice or death, time remains a renewal of the present without things changing.

6

not everything passes with time

Beautiful sky, true sky, look at me changing!

PAUL VALERY

TWENTY-FIVE CENTURIES ago, Parmenides believed that time could not be explained.[1] He saw movement as a succession of fixed positions, so that everything could be described from the single concept of immobility. The future was thus just an illusion, an idle entity on the order of "nonbeing." Carried away by his fierce immobility momentum, Parmenides also rejected the concepts of change and movement, arguing that they contradicted the spontaneous tendency of reason toward identity and permanence. Facing him and standing on exactly the opposite point was Heraclitus, who proposed blending matter and movement. According to him, everything was mobile, so mobile in fact that it was impossible to imagine a fixed point from which to evaluate changes happening in the world, or to explain anything.

Over the past two millennia, these two currents of thought have not stopped fighting each other. Through various thinkers, Being and the Future have been fighting a ruthless war. As far as common opinion goes, Heraclitus seems to be the winner. "With time, everything goes away," we commonly say, adding, in order to be perfectly clear, "and nothing ever lasts forever."[2] The future thus became time's main guise, its shroud. But we should not forget that in addition to philosophy, physics has also been involved in these intellectual duels. And she chose the other side, Parmenides's side.

In fact, physics endeavors to find invariable relationships between phenomena, connections shielded from change. Like Parmenides, physics seems to be fascinated by the invariancy or immobility idea, so much so that when it turns to historical or evolutionary processes, it describes them with forms, laws, and rules that do not depend on time. In this way it hopes to construct a "legislation of metamorphosis" based on notions that are unconstrained by time. The laws it follows are presented a priori as timeless, "exterior" to the universe, as if they were flying high above time. The approach of physics is thus to express the future based on elements that escape the future, to tell stories based on rules that *are* but do not *become.*

Did physics have a choice? Certainly not, because it is

impossible to express the future by referring only to the future. How can one construct a theory from fluctuating concepts? If it were not supposed that the notions appearing in the laws of physics are inflexible, what would happen to the status of these laws? If the concept of movement were itself mobile, what could be firmly said about it? What would a law of gravity be that changed daily? To avoid tackling these embarrassing questions, physics assumes that these laws are invariable through time, even if that means physics has to broaden or transform them whenever some facts happen to come along and refute this a priori assumption. More precisely, it postulates—and this is its radical Parminedesian side—constancy *through time* of the connections between the terms its laws link.

A fundamental theorem gives all its strength to this idea by linking, almost mechanically, two apparently quite distinct notions: preservation and symmetry. In 1918 the mathematician Emmy Noether determined that to any invariance in a group of symmetries is necessarily associated a maintained quantity under any circumstances, that is, a law of conservation. Let's postulate, for example, that the laws of physics are invariant through time translation (they don't change even if we change the moment of reference, the "origin" from which the lengths are measured). This means that the laws governing any physics experiment cannot depend on the particular moment when the experiment is conducted; every moment equals another, so that there is no particular moment that can be an absolute reference for other moments. Applying the Noether theorem, we discover that this invariance in time translation has, as a direct consequence, energy preservation. Let's use an example to back this up. Imagine that gravity changes periodically in time, being, for example, very weak every day at noon and very strong at midnight. We could then bring a heavy load to the top of a building at noon and send it over the edge at midnight. The energy accumulated this way would be more than the energy spent. There would no longer be any energy preservation.

The law of energy preservation thus holds much more meaning than in its usual formulation; it expresses nothing less than the continuity of physical laws, their invariance under time.[3] Under control of the law of energy preservation, time becomes the guardian of the physical world's memory and the base of its future. You have to picture time with a rucksack, enabling it to scrupulously carry physical laws from moment to moment without changing them.

We will certainly object to this way of studying time in which the present universe does not bear much resemblance to the primordial universe. But in fact it is the physical conditions, not the laws, that have changed. In all of its space-time points, the universe keeps a memory of what it has been along with the possibility of replaying the script of its first moments. Thus, when physicists provoke very violent particle collisions in their high-energy accelerators, they obtain information about the universe's very distant past.[4] In fact, they create (or rather, re-create), in a very small volume and for a very brief time, the extreme physical conditions of the primordial universe (very high temperatures and very high energy density). Very large numbers of particles emerge from these terrible collisions, resulting from the mass energy conversion of the incident particles.

Most of these particles do not exist anymore in the universe. Too ephemeral, they quickly transformed themselves into other lighter and more stable particles that constitute matter today. But according to invariable physical laws, the universe secretly kept the possibility that these objects may someday reappear, provided some physicists (helped by taxpayer dollars) pushed it a bit. And this is how the universe allows its old memories to be updated; a violent clash between two particles is just like a bath in a fountain of youth.

It happens, though, that we try to know if physical laws may have changed over time. In practice, this problem comes down to a more economic formulation: we assume that the laws have not changed over time, but the universal constants that they imply—for example, the constant of universal gravitation—have. In the 1930s the physicist Paul Dirac had proposed such a hypothesis with the objective of giving an account of cosmological as well as microscopic phenomena.[5] This theory cannot be applied as such today, because various cosmological observations have established the incredible constancy of physical universal constants over very long periods of time. But very recently, the observation of some absorption lines in quasar spectrums has suggested that a certain dimensionless constancy, said to be "of thin structure" and characteristic of electromagnetic interaction, could have been different in a previous period.[6] To be continued . . .

A surrealist novel by a writer who has long been forgotten gave a brilliant example of the law change theme, but within the social field. In *The Uncertain City* (*La Ville Incertaine*), written during World War II, J. M. A. Paroutaud depicts a city where laws and rules change everyday.[7] Nothing is set in stone except that laws change perpetually. You may have the right to steal one day but not the next, with nobody knowing the laws in use at any moment except for the "cap people," who are in charge of arresting (and then eliminating) the offenders. Every morning a perverse demiurge decides what is legal and what is not, decreeing ephemeral laws in a secretive way; organizing an appalling game of torture in the city's heart, part lottery and part violence; and preventing any way of definitively setting notions of good and bad.

Such a situation is found not only in fiction. When it was under Taliban control, Kabul, Afghanistan, incarnated this kind of "uncertain" city. In December 2001, a Kabul resident explained, "Nobody really knew what was authorized. One day, the religious police would come with their whips and we would then learn that this or that was impious."[8] We can feel the fear inherent in such a situation; when the link between law and permanence is destroyed, all references vanish. Justice, human relationships, and maybe even the meaning of life are deprived of their points of reference.

We better understand what is so ideal and reassuring in physics; it can conceive time only by imagining it converted into invariance. According to physics, time advances by fixedly preserving the shape of the world's laws. These are like its eternal diamonds. Only the physical conditions change in the universe.

Contrary to what we always say, the so-called flight of time is thus neither reckless nor totally destructive. Not everything passes with time.

7

boredom; or, time exposed

The oyster's boredom produces pearls.

JOSÉ BERGAMIN

Vincent Tuquedenne kept on killing time with his heels, stamping on those empty moments which he couldn't even fill with cafés crèmes.

RAYMOND QUENEAU

PHYSICAL TIME IS often presented as an abstraction, as an ethereal reality, inaccessible, impalpable. This is an exaggerated point of view. There is an experience—purely metaphysical—of physical time: we know it as boredom. When nothing happens, when nothing is coming, we experience a sort of empty, more autonomous time shorn of its fancy dress and glitter. It is a nonelastic time, which seems to be dissociated from the future and change. It is time stripped bare, physical time as Newton first defined it.

We are bored when time seems empty or sterile because nothing happens, because we have nothing to do, or because we fail to get interested in what we are doing. We are bored when we are condemned to a wait we cannot shorten. But we also very often get bored when we no longer expect anything. Time then sheds everything it is usually mingled with.

Boredom is like a coin; it has a certain value and two faces. The tails side (which we'll call the bad side) shows signs of a lack of being, an existential vacuity, which reminds us at each moment of our "eter*null*ness," to use Jules Laforgues's coinage; it can turn a smart man into "a walking shadow, a thinking ghost," to speak like Gustave Flaubert.[1] But the heads side—the good side—offers the possibility of an open contact with oneself. Though always presented as pure negativity, hell to escape from, an experience to avoid, boredom is actually able to flip over and become an occasion to learn about oneself. Boredom, then, works wonders that no turmoil can reach, and there lies the miracle of boredom.

First, boredom detoxifies our relationship to time; nothing happens except the passing of time. It puts us in contact with time reduced to the succession of moments, free of what usually contains or contaminates it. In a way, boredom acts as a temporal blowtorch: it burns what is at the periphery of time, clarifies our relationship with it, and makes its skeleton visible. Only the *tick-tock* remains. Boredom throws us into the mushy limits of the present, deriving its richness from this temporal shrinking. By emptying the universe of all consistency, by freeing it from future worry as well as from memory, it prepares something wonderful, "like spreading a white tablecloth for the holidays" and thereby gives us the key experience of being radically at odds with the world.[2] When boredom accompanies us, we are utterly alone, making it therefore impossible to miss oneself. Boredom looks like the exact corollary of self-consciousness and is thus another way to designate it. When it reaches this dimension, it becomes a nourishing void that is able to reveal what is latent in oneself; it peels the time of existence away as we peel wallpaper off a wall. In fact, only boredom gives us the chance to chew on "pure" time, a time that is very close to physical time.

Experiencing boredom, however, is not enough for us to be able to conceive the idea of physical time. When it comes to conceptualizing it or expressing it mathematically, suffice it to say that man knew boredom long before inventing modern physics! But we are so used to the idea of "physical" time (at least in the West) that we no longer perceive its oddness. We even have a hard time understanding what could have been so extraordinary in Galileo's discovery of gravity—what gave time its laurels of being a mathematical variable. Wasn't it obvious that time should be transformed into a quantifiable entity?

Until the sixteenth century, the common idea of time was focused on everyday concerns, and it didn't cross anyone's mind to directly involve time in the expression of a physical law.[3] As a matter of fact, several millennia passed between the first measurements of time passing using gnomons, the first "shade clock," and the first working conceptualization of time.[4] This proves that clocks, whatever kind they are, do not show time explicitly the way physics conceptualizes it, but only as an effect of its passing. Clocks were numerous for a very long time before one man—Galileo—had the idea to spice up time with physical world flavor by giving it an authentic structure: to each moment corresponds a particular value of the time variable, *t,* and every duration is comprised of moments without duration, similar to the way a line is made out of dimensionless points. Formalized in a more rigorous way by Newton, this mathematization accentuated the personification of time that had already been initiated in Greek philosophy.[5]

If you're surprised that it took so long before someone had the bright idea of mathematizing time, you're forgetting that the status of physical time, far from being obvious, is actually very "special." It is a time beyond any differences in subjective judgment, a time supposedly beating in the heart of nature, whether that be right here, next to us, or at the outskirts of the universe! It's also ignoring the fact that numerous human societies have never felt the need to make up the idea of homogeneous time, China being the most famous example. As told by François Jullien, the Chinese knew about calendars and clocks of all kinds, but never conceived time as a monotonous succession of qualitatively identical moments.[6] Instead they see it as an ensemble of eras, seasons, and ages, each with a specific substance and particular attributes, so that there is not one single thread that can really link them together.

Several other examples clearly show that it is definitely possible "to speak of time" without having any homogenous conception of it. The *Orokaiva,* for example, the inhabitants of the northern part of Papua, New Guinea, do not have a word to express time itself, but they have one to say "day," another for "night," others for "before," "now," and "after."[7] Their language also enables them to indicate temporal distance, "depth in the past," thanks to verbal forms and diverse conjugations.

But then how on earth did this idea of physical time appear? It's difficult to describe the different intellectual steps that made its conception possible. Henri Bergson is one of the rare people who dared to give it a try, albeit with a certain naïveté.

Bergson defended the idea that physical time resulted from a simple extension of our subjective experience of duration applied to things. According to him, we ultimately founded a scientific representation of time because we expanded our personal "experience" of time to the world around us, out of simple continuity. I have to consider, explains Bergson, that the temporality of the sugar melting in a glass of water on the table is actually the reflection of my waiting, perhaps of my impatience. By going from my own consciousness to the glass of water, then to the table, and on to other objects around me, I can go from the affirmation "I endure" to the conclusion "the Universe endures," too.[8] "We don't endure in solitude," writes Bergson to signify this personal appropriation of the world by consciousness.[9] Exterior things endure as we do, so that time, considered in this extension, can slowly take on the appearance of a homogenous environment. That's how we go from time as experienced by consciousness to the mathematical variable *t* of the physicists. At the end of this generalization process, the self and the whole eventually merge, or at least join together.

Bergson's thesis is far from gaining general acceptance, partly because making physical time a direct extension of experienced time presupposes that it is close to our subjectivity. Einstein was strongly opposed to this conception. He explained to the philosopher, "It's up to science to tell us the truth about time, as about everything else. And the experience of the perceived world with its obvious facts is just stammering before science's clear words."[10] As a matter of fact, nothing proves that it is possible to link the conventions of common knowledge and the structure of things.

Time, as described by physics, matured so slowly in man's mind because its properties are contrary to our intuition; language and common experience do not have direct access to it, because they only grasp it through everything it makes possible. As quantum physics has even more brilliantly proved, radically breaking with intuitive ideas, intelligence, not intuition, enables us to elaborate the concepts needed to explain physical reality.

8

what makes time pass?

I wander through the days like a
whore in a world without sidewalks.

EMILE CIORAN

In the past there was much more
future than there is now.

THE CAT (A FRENCH COMIC BOOK CHARACTER)

THE FIRST MATHEMATICAL expression of physical time, announced by Galileo and formalized by Newton, consisted in assuming that time had but one dimension. The argument was simple: one figure is enough to date a physical event. Thus there is only one time at any given moment. And since time is always passing, we can represent it with a perfectly continuous line.[1] This representation complies with what our experience tells us: events are superposed in time (meaning they happen at the same time), with no gaps between them. Time knows nothing about coffee breaks and never takes vacations. There is no "breach" in its envelope that would allow the slightest escape, however short. Time is thus compared to a flux composed of infinitely close moments, one succeeding the next.

We are so deeply impregnated with this centuries-old depiction that it has partially anaesthetized us. By force of habit, we think it is enough to nullify the question of how to depict time. Simple formalization of the river picture? Elementary mathematical expression of our intuition about time? When we think about it, this representation raises strange questions that refer to what we might call "timeline problems."

First, to create a line starting from a point, you have to give it what is always missing in an instant to turn it into duration—which is time! The depiction of time as a line is then fundamentally incomplete; it omits indicating how this line is built. Since the present doesn't bring another present by itself, there has to be something, a "small engine," to do this work on its behalf. This little engine that pulls the thread and continuously renews the present, what is it other than the very "heart" of time? Isn't it what prolongs each instant into temporal continuity, into duration? Without its dynamic, the newness of each instant could not arise. This leads us to change the way we look at the timeline: the heart of time exists less in the line through which we represent it than in the hidden dynamic that builds this line.

A second problem appears: in order to say that an

infinite number of points create a line, isn't it necessary for them to coexist *at the same time* in front of our eyes? Bergson had noticed that representing time with a line was actually only a spatialization of time, verging on its own negation:

> If we establish an order in succession, it's because succession becomes simultaneity and projects itself in space . . . To give this argument more rigorous shape, let's imagine a straight, indefinite line, and on this line a moving, material point A. If this point became aware of itself, it would feel itself changing since it is in motion, it would notice a succession. But would this succession appear in the form of a line? Yes, certainly, provided it could rise above the line it is moving on and simultaneously perceive several juxtaposed points. In doing so, it would create the idea of space, and it is in space, and not in duration, that it would see the changes it experiences unfold.[2]

In fact, a line can be perceived as a line only by a spectator observing from an exterior standpoint. Yet "levitating" above time is impossible; we can never extract ourselves from the present to observe its continuity with past or with future. So then how do we manage to speak about the "shape of time" knowing that it implies an exterior view of time, which is precisely what we do not have? Would we be like goldfish, mysteriously able to describe the external shape of their bowl?

In his *Confessions*, Saint Augustine, predicting this difficulty, wonders about being able to feel time passing: "How can I simultaneously be in the present and have enough distance to notice that time is passing?" Almost sixteen centuries later, this question keeps destabilizing the most stable minds, even if the argument put forward by Bergson to contest the spatialization of time doesn't really hold firm anymore. Today we know how to characterize the fact that a line is a line without the necessity of plunging it into a space bigger than itself: its "topology" and its essential properties—its continuity, for example—can be mathematically defined in an intrinsic way, without relying on the line's "exterior."

We can then wonder about the localization of the timeline. If everything is contained in the line, in what space outside time does this timeline have to be drawn? Is it hanging in the void, or does it depend on "something"? We encounter once again the riverbank problem evoked in the river metaphor described in chapter 3. In *what* does time unfold itself? Encompassing everything, how can it be represented *in* something? Does an "outside" of time exist? Either we can imagine that time creates the world as it passes, instant after instant, as if it carried the world on its shoulders as it went along, or we can imagine that it only traverses a territory that is already here, present from time immemorial.

In this way, two radically different interpretations of physical time appear. In the first hypothesis, representing time with a line presupposes the production of this line—as if time created the points it traveled, as if a creative power inherent in the present was pulling it from nothingness, always making a new entity emerge. In a second hypothesis, it is like an infinite scene, already there, waiting for what could happen in it, in which time comes to unfold itself. Which one of these two points of view do we have to choose? And above all, do we have to choose one?

We will leave these questions on hold, since what matters first and foremost is to discuss what shape(s) the line of time has the ability to take.

eternal recurrence; or, the circle's vices

If you push a square crate in front of yourself
with all the difficulties it represents
and there's a kid playing ball next to you,
you'll take notice!!

FERNAND LÉGER

We keep on learning and we'll never stop.
We can't know everything.

KEITH RICHARDS

PLANETS REVOLVE AROUND the sun, days succeed nights, seasons follow one another and look alike, the Tour de France ("the big loop") brings the peloton of cyclists back year after year, our hearts beat as rhythmically as windshield wipers, and tax forms regularly arrive in our mailboxes. From the simple acknowledgment that time makes certain events repeat, that it sometimes reiterates what it authorized itself to produce on one occasion, we have deduced that time itself is cyclical.

This is an age-old reflex. The depiction of time as a circle ruled for centuries. Appearing as the most perfect geometric form, the one without beginning or end, so symmetrical that no one would dare tamper with it, the circle incarnates the very idea of perfection. Doesn't the contemplation of a round shape provide some visual pleasure?[1] And moreover, a circle rolls(!), which is another reason it fascinates us so much. This circle magic goes from the sun to the slightest coin, passing through balloons, pies, soap bubbles, and a woman's curves. In short, with a little imagination we never stop living the fabulous adventure of perpetually winning the local lottery. There is nothing surprising, then, in the fact that the idea of time doing infinite loops could have prevailed in humanity's major myths,[2] in certain religions and some philosophical systems (the Stoics and Pythagoreans, for example, and more recently Schopenhauer or Nietzsche). Roughly speaking, there are two ways to look at eternal recurrence—one of them very comforting, the other not at all. On one hand, we can find it soothing: relativizing all events, including death, creates more serenity than a dramatic sort of time with a unique beginning and a definitive end;[3] it frees the relationship to the past from all sorts of nostalgia such as remorse, regret, repentance; it manages miraculously to achieve two manifestations of happiness, a lasting one and a "here we go again" one. It's understandable that such a view can soothe minds that are upset by the idea that some losses are definitive. But the eternal recurrence can also be depressing; if everything comes down to the same thing, it's because the will has no real effect, acting has no meaning, and freedom has no existence.

Before discussing what physics thinks of this cyclic conception of time, we should recall how various systems of thought have stated it. We will also try to see if their way of functioning does not actually rely on confusion between the repetition of phenomenon *in time* and the repetition of *time itself*.

Let's start with the Stoics, for whom the world perishes before identically regenerating itself, indefinitely, with the same people, in an uninterrupted flow of eclipses and rebirths. What we call the "future" is thus only the return of the past; the effect of time adds nothing to what is or what has been. All newness is impossible, as if the world had taken refuge in itself. It does not open up, everything is given at the starting point; there is neither destiny nor freedom, only necessity. In this obviously repetitive scenario, it is not time but the history of the world that is cyclical.

The same remark applies to the Pythagoreans, who, based on their observations of the celestial revolutions and the seasons' rhythms, conceived the "big year" cycle, at the end of which the whole sky returned to its initial configuration.[4] We find this observation of the revolution of the planets at the origin of cyclic time in Greece, Iran, and India, but we do not know if these are reciprocal influences or autonomous traditions.

Taking a religious example this time, specifically in the Brahman tradition, a cycle is a *yuga* or *yoga,* a "link" between one cosmic period and the period following it.[5] Each *yuga* is preceded by a dawn and followed by a sunset. Since the golden primitive age, there has been a succession of four *yuga,* with uneven and decreasing lengths, from four thousand to ten thousand "divine" years.[6] At each change of era, humanity loses a quarter of its virtue, lifespans grow shorter, mores become more lax, and intelligence wanes. The succession of *yugas* is accompanied by human decadence in the biological, intellectual, moral, and social realms, which continue to deteriorate unless a major cataclysm makes it possible to return to happy times. In this context, coming out of cyclical time would be the equivalent of a release bringing the soul's definitive salvation. But *yugas,* considered to be cycles, are not really cycles at all; they do not have the same length, don't repeat the same thing identically, and can be very limited in number. It is wrong, then, to consider them to be of the eternal recurrence, and even worse to associate them with cyclical time.

Nietzsche's approach is more stimulating. First, because the fiery philosopher, who believed in neither metempsychosis nor reincarnation, looked for what could support the concept of eternal recurrence in the physics of his time, in the area of the statistical thermodynamics.[7] Second, because his argument had a larger impact than simple physics did. He drew a kind of moral code out of them that said if evolution returns upon itself to create a big cycle where everything eternally reappears, then we have to divide our most daily experiences between those that do not deserve to be lived again and those we might wish to see happen again. Rather than suffer the future, you have to desire it. He writes:

> My friends, I am the one who teaches the eternal recurrence. Here, I teach that all things recur eternally, yourselves included, and that you have already been here

> an incalculable number of times and all things with you; I teach that there is a big, long, immense year of evolution, which, once finished, turns immediately back like an hourglass, tirelessly, so that all these years are always equal to themselves, in the smallest and biggest things. And to a person dying I would say, "See, you're dying and will disappear presently . . . But the same power of causation which created you this time will return and create you again . . . for a life absolutely the same as the one you are deciding right now, in the smallest and the biggest things."[8]

The point is to want life as it is: not to be opposed to reality, but to consent to it intensely; despite the rupture of facts, to preserve the continuity of action.

For Schopenhauer, who also depicted time as a circle forever closed upon itself, the concept of the future that is so precious to Hegel is an illusion. Time is always pretending to announce a new end but actually returns to the starting point. It turns but does not progress. Thus, there is no History with a capital "H," since the same small stories repeat themselves infinitely; with joy, expectations, and pain alternating without respite, time is not carrying out its traditional mission, which is to make the future happen. Worse, the past must re-happen. As it says in Ecclesiastes, "The thing that hath been, it is *that* which shall be; and that which is done *is* that which shall be done: and *there is* no new *thing* under the sun."[9] We think time is free and alive when it reveals itself to be frozen for ages. Nothing forbids changes to interfere between two successive cycles, but they are perfectly illusory. And it is because they are illusory that these changes allow humanity to accept the eternal repetition of its history, meaning, according to the happy Schopenhauer, the eternal "repetition of the same drama." Just simply considering them possible is enough to fuel the efficiency of the mirage of the will and the illusion of freedom, when in reality everything returns to the same thing and nothing changes.[10] But Schopenhauer was wrong on one point: time does not run in circles because history repeats. Events can certainly repeat. But what about moments?

The concept of eternal recurrence has had incredible luck. It even became a real philosophical brew to which we owe, among other things, Sisyphus's famous rock and Ixion's lesser-known wheel.[11] Isn't such fertility surprising? As soon as they become associated with the concept of cyclical time, don't the doctrines of eternal recurrence suffer from a certain amount of internal incoherence?

First, in this perspective of cyclical time, each moment of time acquires a double, nearly contradictory, status. As a matter of fact, every instant is peripheral and central: peripheral since it is only a point situated on the circumference of a circle, and central since, being traveled an infinite number of times, it becomes a kind of fixed and eternal point.

Next, taken literally, the idea that a temporal cycle can repeat itself endlessly is paradoxical. Let's assume, however, that it could be possible. There are two ways this could happen: when traveling a given cycle the second time around, we remember what the first trip was like; thus it is not an authentic repetition of the experienced experience in the first cycle, but a simple "revival" of a scenario without surprise, since we no longer discover what we are reliving. The other possibility is that each start of a new cycle "wipes the slate clean," meaning that each cycle is experienced for itself as a unique and new event, regardless of what came before and unaware of what will come after. This case isn't a real recurrence either, since the one living it is ignorant of the fact that he is only reliving it.

To summarize, in order for there to be a future and not only the same old song, chance, unpredictability, and changes have to be working each time so that each cycle is different from its precedent. The difference injected in the repetition prevents repetition from occurring—and we're no longer in eternal recurrence!

What remains of time in the doctrine of cyclical time? Nearly nothing. It eventually implies the negation of the time flow, because it denies what is fundamental: the mutual exclusion of the past, present, and future. Going toward the future, we return to the past and then return once more to the present. We lived and we will live the present we're living. We are living and we will live the past we lived. We are living and we lived the future that we will live. Nothing is really passing then. Everything is always already there; everything is always still there. Time loses all operativeness. Eternal recurrence deploys a sort of "non-time." That's why it was so successful and why Nietzsche himself did not take this doctrine very seriously. Incidentally, he speaks about the "absurd" eternal,[12] aware that the very idea of time depends on the radical differentiation of past, present, and future, an idea that makes sense only if each instant is necessarily new compared with any present that became past.[13]

The idea of eternal recurrence, stated in terms of cyclical time, implicitly relies on a false syllogism: taking note that certain facts repeat themselves, it suggests that this phenomenon of repetition implies that time repeats itself. Yet the existence of cycles in time does not at all mean that time itself is cyclic. Does the fact that there are cyclical phenomena, but also geological, chemical, biological, and psychological phenomena, with their specific temporalities, oblige us to invoke the existence of cyclical time, "geological time," "chemical time," "biological time," and "psychological time"?

At the end of the eighteenth century, a man who has been nearly forgotten today, Jean-Henri-Samuel Formey, had understood that the apparent similarity between time and temporal phenomena was setting a trap for understanding, because nothing proves that time has anything in common with the process whose unfolding it enables. Formey, a member of the Royal Prussian Academy, is quoted by Jean-Jacques Rousseau, author of the article about time in Diderot and d'Alembert's *Encyclopédie*: "Duration is only the order of successive things, as they succeed each other, leaving aside all internal qualities other than simple succession. . . . Time is only an abstract being, which is not endowed with the properties the imagination attributes it with."[14]

Formey, unjustly forgotten, had indeed seen time correctly.

10

causality; or, the impossible *tick-tock*

> God's ways are straight, but malicious souls will stumble on them.
>
> **BLAISE PASCAL**

Time is a very simple being for physicists. Its single and unique dimension gives it much less topology than space, which has three dimensions. As a matter of fact, two, and only two, possible configurations exist for the "shape" of time. The line representing time is either open, or it closes on itself. In the first case it amounts to a straight line. In the second it is equivalent to a circle. There are thus only two types of time possible: linear and cyclical time. What we have called the "time flow" appears on these two types of curves because they are oriented, meaning they are traveled, in a definite direction—from past to future. That's why we often put a little arrow on the curve of time to indicate in which direction it is traveled.

For centuries, the magic of the circle was at work, and the idea of cyclical time prevailed despite the logical difficulties it posed. Then linear time won out. The victory of the straight line over the circle was, according to some historians, an indirect inheritance of primitive Christianity. The invocation of a divine design eventually leading to God's rule on earth has undoubtedly contributed to reinforcing the idea that founding events of a new time can occur. This is in contrast to the cyclical conception, which insists that the same events recur; for an event to be singular, unique, the time flow cannot repeat itself. From this point of view, one of the differences between Judaism and Christianity is that for Judaism, salvation is yet to come, since the Messiah is still expected, whereas for Christianity, the "center of history" is located in the past, in Christ's death and resurrection.

It's through more prosaic, utterly profane arguments that physicists adopted linear rather than cyclical time. Their view is that if time cannot turn in a circle, it is due to an apparently simple principle, the "causality principle." In its classic formulation, this principle was mixed with the idea of strict determinism; as Leibniz wrote, "Nothing is done without sufficient reasoning. That is to say that nothing happens without it being possible for the one who would know matters enough to give a sufficient reason to determine why it is like this and not otherwise."[1] The causality principle is usually expressed by saying that every fact has a cause and that the cause of a phenomenon is necessarily anterior to the phenomenon itself. It's worth noting that, expressed this way, causality has not been the prerogative of the scientific catechism. Almost all philosophers, from Aristotle to Kant via Marechal de La Palice,[2] have founded the basis of their thought on the necessity of a cause, which has sometimes been described as the fundamental shape of our perception of the world, an a priori shaping of our understanding, which always needs to imagine order to avoid becoming hopelessly lost.[3]

Physics endowed this very metaphysical causality principle with its radical nature. So how did it happen to exclude the fact that physical time could be cyclical? In circular time, the future returns on itself so that everything reappears, so that what we call the cause could definitely be the effect and vice versa. The causality principle would thus be unenforceable. Moreover, the circularity of time would lead to some pretty curious situations. If going toward the future were the equivalent of going toward the past, a human being could erase from the past one of the causes that enabled his birth—preventing a meeting between his mother and father, for example. Such a paradox, possible with cyclical time, is not possible with linear time, which arranges events according to an irreversible chronological sequence.

However, how the principle of causality is expressed has evolved considerably through time. After playing a major part in seventeenth- and eighteenth-century physics, the cause concept saw its importance decline in the nineteenth century with the appearance of probability used in statistical physics.[4] In the twentieth century, quantum physics gave it the deathblow. As a matter of fact, when it comes to quantum processes, the way the physics of the infinitely small uses probabilities forbids speaking of cause in the strict sense of the word. The physicist Max Born was the first to note this. In 1926, when he was studying, from a theoretical point of view, the way an electron changes when it is projected at an obstacle—an atom, for example—he discovered that the "wave function" of the electron, at first a simple flat wave, slowly changes, becoming distorted as it nears the obstacle, and eventually diverges to spread in all directions.[5] Yet the corresponding experiment existed, and with perfectly clear results: if we detect the electron on a fluorescent screen, we do not see a diffuse light spreading over the entire screen wherever the wave function is supposed to be present; on the contrary, the electron shows up at one single point on the screen, where a scintillation indicates the impact. When the experiment is repeated with other electrons, the same phenomenon happens again, except that the impact point changes on the screen; it seems to change randomly. Max Born concluded that the wave function does not control the exact movement of the particle but only the probability that it could be detected at one point or another on the screen.

Physicists noted this revolution and reworded the principle of causality to no longer refer directly to the idea of cause but simply to an obligatory and absolute order between different types of phenomena, without describing one of them as the cause of another.[6] In their formalism, causality is nothing more than a method of ordering events, a "rule" that positions events according to a strict order. Pared down in this way, the principle of causality simply states that time is not whimsical, that it flows in a determinate direction, so that it is always possible to establish a well-defined chronology if events are causally linked.[7] There cannot be any reversal; though time may go *tick-tock,* it never goes *tock-tick.* As we will see later on, this simple idea that time does not walk backward—and that eventually that is what characterizes it—has significant (and sometimes disconcerting) consequences. For example, it has led to the prediction of the existence of antimatter!

The other side of the principle of causality consists in saying, as the expression goes, that "the same causes produce the same effects." Time can thus seem to have *ticks* along its *tick-tock;* some phenomena reproduce themselves exactly the same, as soon as their causes repeat themselves. By making possible the identical reappearance of some events at different moments in linear time, causality organizes the repetition of some events and enables history to "serve the same dish over again."

A short story from Thierry Jonquet, "La Vigie," is a spectacular example of the invariability of certain effects produced by certain causes.[8] An old man who tirelessly fought during World War I uses binoculars from his house on high to observe the daily lives of his neighbors in the surrounding buildings. Day after day, he recognizes tragic situations he had already experienced some seventy years earlier: a young mother prostitutes herself at home, not noticing that her daughter sometimes witnesses her actions; a doctor who is exhausted from unceasingly healing destitute people becomes suicidal; a father of a large family is fired, starts drinking, and beats up his children, whom he cannot feed anymore; young Muslims, fed up with being humiliated on a daily basis, prepare their revenge by buying fuel jars and nails to create bombs. The old man knew all these situations, under almost identical appearances, when he was on the front or on leave, and they all came to tragic ends: suicides, murders, or terrorist attacks. He senses that the experiences his neighbors are living will lead to the same disasters. He writes several letters of warning to the mayor of his town but gets no reply. The causes he has identified are then free to produce their effects—suicides, murders, and attacks—but this time the effects are concentrated in a single day. "I say that there's no such thing as chance," grumbles a bar owner when told that more than twenty people died the same day in the projects.[9]

Causality, by authorizing "temporal cloning" of some chains of events, is actually a guarantee of systematic "here we go again!" It is even thanks to this that we notice it. Causality creates two major consequences that are only apparently contradictory: on the one hand, it prevents time itself from being cyclical; on the other hand, it guarantees that events can repeat themselves in time, which implies cycles in time. In sum, it authorizes the repetition of phenomena while forbidding repetition to time.

As soon as the linearity of time was affirmed, new perspectives opened. Marked by unique events, stretching toward a necessarily new future, it broke radically with the stuttering of circular time and its monotonous iterations. It transformed the future into an adventure. Before, it was the old same song, the sempiternal and nothing else.

With linearity came historical production, invention, the new. But also some cycles. And sometimes the irreparable, the definitive.

By design, linear time rushes straight ahead of itself. Each day it makes is a new day. The straight succession of its *tick-tocks* nibbles away bits of the circular perfection. It gives us room to maneuver and an appearance of freedom.

11

"time travels" and other uchronias

I flew too far into the future;
A shiver of horror overcame me.
And when I was looking around myself,
time was my single contemporary!

FRIEDRICH NIETZSCHE

I had lost the sense of history, as
happens in a lot of diseases.

MAURICE BLANCHOT

EACH ONE OF us has experienced time as a jail without bars, a jail we would like to leave so we could leisurely wander the axis of time, so we could, in short, "time travel."

Would time travel be about recalling a lost past? About reliving happy moments over and over? Finding lost friends and relatives? Changing eras without changing age? Changing age without changing eras? Observing the past and future on some sort of movie screen while living a kind of "temporal teleportation" that would divide one's personal time from historical time? Or would it mean going back in the past to transform historical reality, to change what has been written or lived, or to give some type of virginity back to the world?

Science fiction authors, who always have been great explorers, have set up many of these various possibilities. With *The Time Machine,* published in 1895, H. G. Wells uses the time travel theme to expound his ideas about the future of humanity. Lyon Sprague de Camp, in *Lest Darkness Fall* (1939), writes about how a time traveler managed to give the Romans inventions they did not have during the first barbarian invasions, such as the steam engine and Arabic numbers, thus increasing their military power. The course of history is of course changed (the Middle Ages are avoided, for example). In *Vintage Season* (1946), Henry Kuttner and Catherine Moore explain the adventures of time travelers who "land" in such a distant future that they become totally insensitive to the suffering of the people they meet. The insurmountable cultural divide creates a kind of emotional indifference. Poul Anderson's Time Patrol series envisages the existence of the Danielians, descendants of the human race living a million years in the future, who don't want anything to perturb their present world, which they consider to be the best of all possible worlds. They create a "time patrol" that is in charge of preventing any attempt to change the past that could, by causal rebound, change their present and their future. As keepers of the fixity of time, the time patrol is in some way the armed wing of causality.

Ken Grimwood toys with the possibility that we could endlessly redo our life. The hero of his 1987 book, *Replay*, forty-three-year-old Jeff Winston, dies of a heart attack on October 18, 1988, and wakes up in 1963, at the age of eighteen, in his former university dorm room in Atlanta. Will he have the same future? No, because his memories are intact and he thinks he can use them to do better than the first time around. Knowing the results of the horse races in advance, he knows what to bet and wins each time. He becomes very rich very quickly, but not much happier, until his second death, on the same date as his first death. When he returns again in 1963, he chooses to experiment with sex and drugs and ends up falling in love with a woman who, like him, is replaying her existence until their third death. Then a fourth life starts, followed by a fifth, and a sixth. In his subsequent lives he always finds his beloved again, but always at a later point, so that each new life is always more of a nightmare than the previous one. At last, death—the real, definitive one—releases him from the exhausting refrain of his successive unhappy lives, none of which has been lived as a "real" life.

In the award-winning book *Timescape* (1980), novelist Gregory Benford uses "tachyons," particles that supposedly allow travel back in time, to explain how Gordon Bernstein, his main character from the present, is able to warn scientists in the past about the disasters their discoveries will eventually create.[1]

Finally, we cite Jean-Marie Poiré's 1993 film, *Les Visiteurs* (The Visitors), as well as *Back to the Future* (1985) by Robert Zemeckis, or to go back even further in the past (which is metaphorically possible with a good memory or good archives), René Clair's 1943 film, *It Happened Tomorrow.* In this movie, a young journalist has the diabolical ability to receive a copy of the newspaper with the next day's date twenty-four hours before the events actually happen. At first, he uses this to win at the horses, but his expression quickly changes when he reads headlines of his own death in a New York bank hold-up he himself had organized. It's a great screenplay that makes the impossible happen. It provides a happy ending, since the journalist does not die, and it saves the newspaper's honor, since the prediction announced on the front page turns out to be correct, except for a very small detail: the mistaken identity of the dead man, due to a subtle episode we will not reveal here.

Time travel comes in an infinite variety of forms, especially when it enables the past, and consequently, the present, to be changed. But in all of these stories the point for the author is not really to put logic to the test. Some inconsistencies remain that don't really bother the physicist or logician who sleeps within us. For example, doesn't the very idea of time travel imply a kind of absurd gap between the time of the traveler and the exterior time in which he is traveling? Doesn't it implicitly imply two different superimposed times in the same world: the temporal traveler's time and the universe's time? Or (what amounts to the same thing) that there are simultaneous states of the universe existing in different times, as the philosopher Alain keenly noted about Wells's *The Time Machine*?

Every good time traveler should therefore have two watches: one showing his hour, and the other one showing the time in his environment of the "moment." Then, if we maintain that physical time is unique and not cyclical, doesn't it logically follow that it is something in which we cannot travel?

Despite these challenges, regular headlines trumpet that physicists are making advances in the field of time travel, and that soon such fantastic machines will be available, or at least conceivable. Do the latest developments in physics provide any hope?

First, let’s ask ourselves this naïve question: if somebody in the future built a time machine, how can we explain that we do not have one already? Let’s say that one of these machines will be built in 2050. It would need to go back only a few years to reach us. It should be able to handle this temporal ride, since that is its precise function. Then why isn’t it already here? Shouldn’t a time machine, able to visit any era, be timeless by nature?

12

antimatter; or, the end of the trip

We are not faring for the
love of faring, that I know of . . .

SAMUEL BECKETT

WHEN YOU ASK a physicist if time travel is possible, she usually answers no and looks irritated. Or she might conceivably give you a glimmer of hope by leaning on general relativity as expressed by Einstein in 1915. As a matter of fact, if you are looking for situations allowing you—in principle—to change the present, this theory of gravitation seems to be the most promising ticket, thanks to, among other things, some "topological tricks."

To understand, you have to recall the essential teachings of general relativity, born from the observation that Newton's theory of gravity was built on a strange hypothesis: the gravitational effect between two bodies, expressed through mutual forces of attraction, was supposed to spread instantaneously through space. If one of the bodies changed shape, the other one was immediately "informed," even if they were several light-years away from each other. This posed a serious conflict with the special theory of relativity, expressed by Einstein in 1905, which stipulated that information could not be instantaneously propagated. Physics lost its coherence. To give it back some order, Einstein had to change the gravitational concept as Newton and his successors had viewed it. In order to do so, he used the research of Nikolay Lobachevsky and Bernhard Riemann, who, a half-century earlier, were working on spaces that had a different geometry than our familiar (Euclidian) space.

Their studies suggested, in a basic way, that gravity might not be an actual force, but rather a local manifestation of the curvature of space. Einstein gave this idea its full force by postulating that the universe's geometry—which was flat, according to special relativity—was curved from the matter it contains, and, in return, this geometry directly determined (that is, without any force coming into play) the movement of physical objects in it. Thus the movement of the earth around the sun no longer resulted from Newton's instantaneous action, but was guided along a trajectory determined by the sun's massive presence. To put it plainly, according to general relativity, the curvature (which cannot speak) "tells" matter how to move, and matter (though mute too) "tells" geometry how to bend.[1]

How about time travel in general relativity? Actually, this question has always been controversial. As of 1937, a Scottish physician by the name of Willem Van Stockum discovered a solution to general relativity equations showing that an infinitely long, rapidly rotating cylinder works like a kind of time machine. But since there is nothing in nature that is indefinitely long, it's doubtful that such a machine could exist. In 1949 Kurt Gödel found another solution to the general relativity equations, describing a rotating, though not expanding, universe in which we can go back in time simply by going far enough from the earth to return to it later. The problem, for those who would like to believe this theory, is that our universe does not turn (or not much anyway) and that it is obviously expanding.[2] In 1976 Franck Tipler demonstrated that in order to create a machine for time travel in a region of space, there has to be matter that is radically different from ordinary matter—and for this reason called "exotic"—that enters into the composition of the machine.[3] But what would this exotic matter be made of? Nobody knows.

Today, physicists speak a great deal about the possibilities that could create "wormholes." Despite their strangeness, wormholes are not pure inventions pulled from the hats of science fiction writers. Mathematically discovered in 1916 by Ludwig Flamm, they are, in a way, shortcuts in the space-time topology allowing a link between two areas far away from each other. A wormhole has two entrances, which can be several millions of light-years apart, but they are linked by a "tunnel" in space-time, which can be much shorter. Numerous physicists, including Kip Thorne, Igor Novikov, and John Friedman, have studied how such wormholes could be used to travel in time: going through one of these tunnels would be enough to travel the millions of light-years separating the two entrances in fractions of a second without having to surpass light speed—in other words, without breaking the laws of the theory of relativity.[4] But this theoretical possibility, used in 1985 by Carl Sagan in his novel *Contact,* is killed in infancy by the fact that wormholes (if they exist other than as pure mathematical theory) are fundamentally unstable; as soon as they form, the tunnel would be destroyed by the smallest entering particle or light radiance. Stephen Hawking reached this conclusion in a principle called the "chronological protection conjecture," which states that time machines will instantaneously self-destruct if we try to build them. They are thus impossible, at least in fact.[5]

In summary, contemporary physics says more or less the same thing as everyday experience, which is that time travel is certainly chimerical. A terse word from Arthur Rimbaud puts it much better than a long, calculation-filled tirade: "You cannot get away."[6]

The physicist's argument for saying "you cannot get away" is the same as the one used to argue that time cannot be cyclical. As a matter of fact, the very wording of the causality principle prevents time travel from the get-go. Time travel, for example, would allow one to go back in time to change a sequence of events that had already occurred, thus having a retroactive effect on a cause that has already produced its effects. Causality claims that there is only one, noncyclical time and that the order in which causally linked phenomenon happen is not arbitrary.

Thus the world becomes a safe place for historians; there can be only one chronology. The fact that an event has happened—really happened—cannot be questioned. It will always be "true" that it happened, even if no memory has registered it, even if it has left no trace, even if it is denied after the fact.[7] Seen from this angle, the past becomes an impregnable fortress.

But does this mean we have to believe in the causality principle? Let's first note that the formal expression of the principle is not uniform. In classical physics it simply amounts to assuming that the shape of time is linear and that no one goes back to the past by going toward the future. But under special relativity, the classic visions of space and time are shaken up, as we will see in the next chapter; neither length nor duration are absolute quantities, that is, independent from the system of reference in which they are measured. Space and time therefore appear intimately linked with each other. They have to be considered together in the heart of "space-time." But in this welding of time and space, how can we integrate the concept of causality? By claiming the impossibility of transmitting energy or information faster than the speed of light in empty space. Time travel and reversals in chronology between causally linked events are thus categorically prevented.

The story becomes more delicate in particle physics, where it is a matter of describing objects that are at once minuscule (subatomic) and swift. Its formalism, the game of equations on which it is based, must manage to mix quantum physics, which deals with very small objects, and the theory of relativity, which deals with very quick objects (ones whose speed is not negligible compared to the speed of light). But if we are not careful, the equations we get might lead us to think up situations in which the disappearance of a particle can precede its appearance!

Accepting such situations comes down to flouting the flow of time, or even denying its existence. We can avoid such situations, on paper at least, by adding some extra mathematical laws to the formalism, certain "constraints" that guarantee that the creation of a particle necessarily precedes its annihilation.[8]

These rules are somehow the armed wing of the causality principle, since they impose respect for a very definite chronology between two types of causally linked events. And that's where the whole story starts: the calculations show that the new "constraints" make the existence of new particles necessary (on paper at least). And unlike all known particles, these new particles have a negative charge. Yet since a negative resting mass corresponds to each negative energy particle, under the action of a force—including gravity—it would move away from an ordinary positive energy. In a way it would seem to be "going back in time,"[9] but this would only be a mathematical term, because we can definitely reinterpret these energy particles as positive energy antiparticles that follow the normal flow of time.

Thus, by maintaining strict adherence to the causality principle, Paul Dirac predicted the existence of antiparticles in the 1930s.[10] The beauty of the story comes from the fact that this astonishing prediction was rapidly confirmed through experimentation; Carl Anderson, in 1932, detected some fifteen antielectrons (also called positrons) among the particles produced by the impact of cosmic rays on the atmosphere.[11] This discovery demonstrated that causality was not a theoretician's fad! Later on, in the mid-fifties the first antiprotons and antineutrons were produced, thanks to powerful accelerators. Composed of what we call antimatter, antiparticles are today well known. Once created, they follow the normal flow of time like everyone else, just as physicists had predicted.

In summary, the existence of antimatter, as shown today, is the material proof (or, more precisely, "antimaterial" proof) of the existence of time, which is a unique direction that orders events in accordance with the causality principle. The appearance of negative energies in the equations ultimately showed nothing other than the impossibility of traveling in time.

On a final note, we very often associate time with space because we are in both, because we cannot extract ourselves from either and are in a way imprisoned in them. Both concepts seem to us without exteriority. However, there is an essential difference between them that causality expresses: on the one hand, by orienting the flow of time, and on the other hand, by forbidding the simultaneous propagation of signals in space. In short, we can move as we like in space, while we cannot willfully change our position in time. Thus, space appears as the zone of our freedom that we can pace as we like, while time is like an embrace toward which we can only be passive; we are "embarked" as Pascal said. This has two important consequences. The first is philosophical: our freedom, if it exists, is not as light as grace, because we are irremissibly chained to the present. The second consequence deals with transportation: any distance covered in space is necessarily time-consuming. Nothing moves in no time at all. In short, a round-trip in space is always a one-way trip in time.

13

1905: "now" says good-bye to the universe

Madame is late. That means she's coming.

SACHA GUITRY

The most beautiful flowers have lost their fragrance.

GÉRARD DE NERVAL

If we often attribute to time properties it does not have, physics links it with concepts that we think are intrinsically different from it. That's the ecumenical side of physics; it happens to unify categories that are split apart by words. That's how, one fine day in 1905, thanks to Albert Einstein, it brought time closer to space and vice versa. This new alliance was immediately perceived as a total revolution, and several biases were its victims. The concept of simultaneity, for example, until then considered absolute, suddenly lost its acclaim.

The simultaneity concept is certainly an effect of the French school system: as soon as the question of physical time arises, an absolute universal time that is the same everywhere and flows identically throughout the universe comes to mind. This time, independent from space, autonomous from physical phenomenon, indifferent to movement, is what we call Newtonian time. It has the advantage of giving the word *now* a perfectly clear and distinct meaning: what is happening *now* for me is also happening *now* for all other observers of the universe. In other words, the simultaneity concept is absolute: at every moment, two observers can synchronize their watches, and the two watches will remain synchronized at any later moment, no matter the movements or speeds of the two observers, since they both remain in phase with Newtonian time.[1] Two events that appear to be simultaneous in the eyes of one observer will be so for all others as well.

But what does Einstein demonstrate in 1905? That physical time is not Newtonian and that everything about it has to be forgotten. By combining time and space in a quasi-conjugal way, he breaks the autonomy of each and changes their properties. This does not mean that space became a new garment for time. More exactly, it acquired the status of a partner, as if both time and space were swimming in the same ontological bath. The simultaneity concept never recovered from this association.

Some history will help us understand what was at stake. At the end of the nineteenth century, physics was leaning on two pillars: Newtonian mechanics and Maxwell's electromagnetism. These two theories seemed accurate in their own fields, but it was quickly understood that their respective principles were contradictory.

Mechanics is founded on the relativity principle, first expressed by the inevitable Galileo. According to this principle, things in a plane flying at cruising speed (Galileo, of course, referred to a boat rather than a plane) happen in the same way as they do on the ground, when the plane is not moving. If a stewardess lets a glass of water slip from her hands, it falls in exactly the same way on the plane as it would if the incident happened in a diner. More generally, no physics experiment could help us determine if we are in a flying plane or a plane on the ground.[2] The movement of the plane is then "like nothing," to use Galileo's words. The consequence of this relativity principle is that nothing is immobile in the absolute: everything moves.

The theory of electromagnetism explains that light is composed of waves. And a wave, as conceived by nineteenth-century physics, is a phenomenon that moves by vibrating "something" from the center of the medium in which it spreads. A wave, for example, vibrates in the water or, more precisely, on the water's surface. With regard to light, it was believed in the nineteenth century that the "ether" was vibrating. People imagined then that the universe was filled with ether, whose existence was necessary for light to spread. What was this substance made of? Was it heavy, solid, liquid, elastic? The electromagnetic theory postulated only vague answers; ether is certainly colorless and probably weightless. Actually, through the years, ether was stripped of most of the physical properties that it was first granted and kept only one: absolute immobility. This conclusion clashed with the relativity principle, founder of mechanics, the other pillar of physics.

Hence the dilemma: either we took the electromagnetic theory seriously and consequently abandoned the relativity principle, or we took the relativity principle seriously and abandoned the idea of ether. Einstein chose the second option. He starts by declaring the death of ether: the propagation of light in no way results from vibrations in a medium; it happens in a void, and nothing vibrates along its way, only itself.[3] He then postulates that the speed of light is identical no matter what the speed of its source or its observer is. When a car approaches me with its headlights on, the light it sends out travels at the same speed in relation to me as if this car were stationary.

This last postulate imposes a change in the way of expressing the relativity principle in order to guarantee that the speed of light is invariable when the reference point changes. From one Galilean referential to the next, the coordinates of space and time do not change the same way: time partly transforms into space, and space partly transforms into time.[4] The frontier that allows us to distinguish them is then dependent on the speed of light from the referential in which we happen to be. We thus have to speak about "space-time" rather than about "space" and "time." This does not mean that length and duration are similar entities, but that they both turn out to be relative to the referential by which they are measured.

By combining with space, time loses its Newtonian autonomy and ideality. It actually starts to depend on dynamics. The consequences are not only philosophical or intangible; the rhythm of any clock will appear to slow down to any observer who does not accompany it in its movement. This slowing down of clocks expresses the relative elasticity of durations, meaning their own relativity. We observe it commonly among unstable particles—muons, for example.[5] Their "own" life span, measured when we remain immobile in relation to them, amounts to a few microseconds (after that, they disintegrate into other lighter particles). But a muon's measured life span coincides with this duration only if it is born and dies in the very same spot in space, which means only if it is immobile compared to the observer. Otherwise, its life span is increased by a factor that depends on its energy or, if we prefer, by its speed compared to that of the observer: the quicker it goes, the longer it lasts, to the point that if its speed is close to that of light in a vacuum, it has all the leisure to appear for a duration far superior to its proper life span.

Another consequence is that the notion of simultaneity, which Newtonian physics clearly proves, stops being absolute. This result comes directly from the invariance principle of the speed of light. Let's imagine we're on board a spaceship and we light a bulb located in the middle of the cockpit. Since the light travels at the same speed in all directions, it arrives at exactly the same time on all sides of the cockpit. Let's imagine now that an observer sees our spaceship approaching him at high speed; since we are going toward him, the light from our bulb, with its invariant speed, will travel a shorter distance in his eyes to reach the rear wall of the cockpit and a greater distance to reach the front wall, so that the observer will first see the light reach the rear wall, and then the front wall. Two simultaneous events for us will not be so for him. An observer who would pass by our spaceship would see things in the reverse order; for him the light would first strike the front wall, and then the rear one. This inversion in chronology does not in any way violate causality, because the two events at stake are not causally linked; being at the same time distant and simultaneous in the cockpit referential, no finite speed signal could put them into relation.

Generally, what is present for us at one instant exists no more or not yet for an observer who is moving in relation to us. It thus becomes impossible to define a "present instant" where all phenomena that happen at the same time in the entire universe would manifest themselves. The nice word *now* is thus deprived of any absolute meaning.

Special relativity also imposes that no observed object can be our contemporary. Sunlight needs eight minutes to reach us, light from more distant stars travels for several years, and light from a very faraway galaxy travels for up to several billion years before we see it. We call this temporal gap "backward-looking time." Looking deep into space is to look deep into the past and to observe slices of the universe as ancient as they are distant. This distant galaxy, observed very young, discovered just after its birth, appears to us as it was several billion years ago; one that is nearer appears the way it was after billions of years of evolution. These two objects, which seem contemporary, are in fact "grasped" at different ages. When we look at them, we can no longer say that we are seeing the "present" of the universe.

For all these reasons we can no longer speak about the universe as if it were a kind of universal metronome. There are now as many fundamental clocks as there are uniformly moving objects. We cannot synchronize them in a perennial way; of course we can adjust their dials at a particular moment, but the hours indicated will stop coinciding several moments later. Each observer will see that the duration indicated on any clock other than his own will always be dilated.

But the causality principle is still being respected, because if for an observer, event *A* is anterior to event *B*, and if a light signal has time enough to go from *A* and reach *B* (which means that *A* and *B* are causally linked), then it is the same for any observer: event *A* will always precede event *B* in all the equivalent referentials. The durations become relative, but the notions of past and future keep an absolute character; from one Galilean referential to another, we change the time interval between two events, but we never reverse their order as soon as they are causally linked. To do so we would have to surpass light speed, which is exactly what the theory of relativity forbids.

does the future already exist in the future?

The future is inevitable, but it cannot happen.
God pays attention to the intervals.
JORGE LUIS BORGES

He who speaks about the future is a rascal,
it's the present that counts.
Invoking posterity is like making
a speech to maggots.
CÉLINE

The future does not yet exist, therefore it does not exist—this was Aristotle's implacable conclusion. But we speak about the future as if it would happen with certainty, as if it were somehow present, as if we were sure that later there will still be some present, reserving our uncertainties and interrogations not over whether the future will exist or not, but over the question of knowing what it will consist of and what will happen. That's where the future's ambiguity comes from: nothing prevents us from picturing it as the fulfillment of all our projects, even the most far-fetched. It does not seem to offer any resistance to our will or our desire, except that no one future path is certain, and we could all die in the next second without any warning. The status of the future is thus particularly ambivalent: certain in its existence, uncertain in its shape.

But where does the future take place? Saint Augustine gave a very convincing answer to this question: the future can exist for us only in our soul—or in consciousness, as we would say today—which is the single entity to have the capacity (along with dreams?) to imagine what is not, especially what is not yet. To take shape, the idea of the future actually implies the idea of waiting, since duration divides us from it; it also implies imagination, since we can anticipate it only in a fictitious way; it implies memory, the only thing able to recognize what will necessarily be repetitive in the future, like fall, winter, and summer, or happiness, sorrow, and happiness again. Memory "furnishes" the future a priori. Without it we could only think of it as a big hole.

It seems to be commonly accepted that the future exists only for the mind, not in and of itself; it exists because we wait for it, and not because it is linked to the present or the past by necessity, by the concatenation of an anteriority that would determine it. But declaring that the future exists only in consciousness, and not in the world, is to grant it a very special ontology; the future would actually only be "the imaginary correlate of a waiting consciousness," as André Comte-Sponville aptly puts it.[1]

It happened—and still happens—that some physicists, inspired by Einstein's relativity, see things differently. According to them, the past, present, and future have "already been here," indistinctly linked in a kind of timeless reality, so that the universe does not have a history of its own. But we observers grant it one because we unfold the thread of time ourselves. Einstein's close friend Hermann Weyl had this point of view: "The objective world simply *is;* it does not *happen*. It is only for my consciousness, crawling along the edge of my bodily universe, that a section of this world comes to life in space like a transient picture, which constantly changes in time."[2] We might actually be the producers of a story that the universe would not have without us; the world would not pass by, but we would make it pass by by passing through it. Everything would then already have been there—the past, the present, and the future—but because of our own path, we would only discover this reality temporally unfolding step-by-step, second after second. We would be the "little engine" of time!

Today, physicist Thibault Damour, a specialist in general relativity, has developed similar ideas, but in his own peculiar way. According to him, the fact that time passes is only an illusion due to the irreversible character of how we commit things to memory: "Like the notion of temperature has no meaning if we consider it to be a system of few particles, likewise it is probable that the notion of time passing means something only for certain complex systems, which evolve out of the thermodynamic balance, and which handle the accumulated information in their memory in a certain way."[3] Time would only be a psychological appearance, linked to the very complex structure of our brain; in the space-time field we are observing, we have the feeling that time passes "from the bottom to the top" of space-time, but in reality it builds a rigid block without any internal dynamic.

Would it then be—seriously—possible that *we* are the engines of time? This thesis, be it founded on general relativity or on a philosophical idealism, is as difficult to accept as it is to exclude. It is therefore best to put forward, against all dogmatism, that such a concept remains simply a matter of point of view.

is time an opportunist?

The interest we have for time comes from
our snobbery for the irreparable.

CIORAN

The disciple: Take a dragonfly, tear off its wings,
it is a pepper.
The master: No, take a pepper, give it wings,
it is a dragonfly.

ZEN SAYING

IN 1929 THE British physicist Arthur Eddington attributed a strange symbol to time—the arrow, which until then had been attributed in mythology to Eros, the god of love, portrayed as a big-bottomed, winged child who stung hearts with his sharp arrows. For physicists, this "arrow of time" no longer represents amorous desire, but rather the idea that it is impossible to modify how certain things happen. "Arrow of time" became the common phrase for metaphorically expressing the "irreversibility" of some physical phenomena.

Although the definition of the expression is free of ambiguity, everyday language introduces some: we very often mix "arrow of time" with "flow of time." Yet these are two very different things. The *flow of time* is a matter of causality, because of the fact that time flows in a single direction, and never goes backward. The *arrow of time* implies the existence of a well-established flow of time, in which some phenomena are themselves temporally oriented, meaning they are irreversible: once accomplished, it is impossible to cancel the effects they produced.

Let's be clear: it is not out of the question that the flow of time and the arrow of time come from the same reality, more profound than they are; that they both are by-products of underlying phenomena that a "new physics" might reveal, but for the time being we have to formally distinguish them.[1]

If the flow of time is a characteristic of time—its essential skeleton, so to speak—the arrow of time is merely a property that physical phenomena do or do not have. Reversible phenomena are said to have no temporal time arrow, and the others are said to be "arrowed."

We have seen how physics represents the flow of time, the link it establishes between itself and antimatter. What does it say about the misnamed arrow of time that would be better called the "arrow of phenomena"?

Since Newton, physicists have wondered if physical phenomena can or cannot unfold in "two directions"; reaching a certain final level, can they return to their initial state? The point isn't at all to know if we can return to the past, but to determine whether or not physical laws authorize physical systems to regain later on—in the future, their own future—the state they once knew in the past.

Picture two balls colliding on a billiard table. After the impact, the two balls go in opposite directions. If friction is insignificant, their speed will remain constant. Now let's film the collision and project the film backward, which is the same as exchanging the respective roles of past and future. What we now see on the screen is another collision of the two balls in question, corresponding to what actually happened, but this time the balls' speeds are reversed.

A spectator who would observe only the reversed movie would not be able to tell if what he were seeing really happened or if the film were indeed projected backward. The reason for this lack of certitude is due to the fact that the second collision is ruled by the same laws of dynamics as the first. It is therefore as physical as the first collision, in the sense that it is just as feasible. Such a collision is thus "reversible" in that its dynamic does not depend on the orientation of the flow of time. This does not at all mean that the colliding balls travel through time, but that for them the flow of time is arbitrary. We could call the past the future and vice versa, without in any way changing the physical process in which they participate.

According to equations of contemporary physics, this statement is also true for all phenomena happening at a microscopic level: they are reversible; they can happen in one direction as well as in the other. Hence the problem of the "arrow of time," because at our scale we observe only irreversible, therefore, arrowed phenomena. If we film any scene from everyday life and then project the scene backward, we can tell from the first images that the film has been reversed. This comes from the fact that at the macroscopic level, in general, we cannot redo what has been undone, nor undo what has been done.

Here we are facing an enigma: how can we explain the emergence of the irreversibility observed at the macroscopic level with physical laws that ignore it at the microscopic level?[2] This question, whose stakes have never stopped evolving, has been fervently discussed for almost two centuries.

The oldest explanation is based on the second principle of thermodynamics, according to which any physical system usually evolves without returning to its initial configuration. Lukewarm water never turns into hot water sometimes and cold water at other times. How did we reach this result? At the beginning of the nineteenth century, Sadi Carnot demonstrated that the transformation of heat into mechanical energy was limited by the one-way direction along which heat transfers occurred, always from warm to cold. We then understood that heat had a special quality that could be linked with irreversibility.

In his *Reflections on the Motive Power of Fire,* published in 1824, Carnot set forth the beginnings of the second principle of thermodynamics, which Rudolf Clausius restated more rigorously in 1865. First of all, this law postulates, for any physical system, the existence of a magnitude called entropy, which is fixed by the physical state of the system. It goes on to point out that the amount of entropy contained in an isolated system cannot increase except by some physical event. Because the total entropy of a sugar cube and an unsweetened cup of coffee is inferior to the entropy of a sweetened cup of coffee, the sugar cube has no choice: it has to dissolve in the coffee. This is an irreversible phenomenon: the sugar melting in the cup of coffee will never get back its parallelepipedal shape or its whiteness. The second principle seems to be able to solve the problem of the arrow of time. But let's not be fooled by appearances.

Compared with microscopic equations, which are all reversible, macroscopic equations typify a more global behavior of matter and are always irreversible. Thus, Joseph Fourier's heat equation established in 1811 indicates that heat always flows from warm to cold, and not the reverse. But if we admit that overall behavior is nothing but the collection of a great number of elementary events, macroscopic equations should be able to be deduced from microscopic equations. However, some are reversible and others are not. How can all of this be tidied up?

Wanting to delve deeper into this question, Ludwig Boltzmann tried to find a link between Newtonian mechanics and the second principle of thermodynamics. Since it is impossible to rigorously integrate the behavior of a very large number of particles, Boltzmann turned to the laws of statistics, giving up the explicit calculations of trajectories for probabilities. He noticed in 1872 that it was possible to create a mathematical magnitude—the function of the positions and the speeds of gas molecules—with a remarkable property: under the influence of molecular collisions, this magnitude could only diminish as it evolved toward equilibrium (or remain constant if the gas was already at equilibrium). It was exactly the analogy of entropy. Thus, the statistical aggregation of reversible equations of particle dynamics leads to an irreversible macroscopic equation. Boltzmann deduced that the growth of the entropy of an isolated system simply expressed the average tendency, manifested through this system, to evolve toward more and more probable states at the molecular level.

But these calculations, which manage to "create" irreversibility from equations that do not have any, make it sound like that this irreversibility is merely an appearance proper to macroscopic systems, an "emerging" property of phenomena that put into play a great number of particles. In short, there would be irreversibility de facto but not on principle. That is, by the way, the meaning of the "repetition theorem" elucidated in 1889 by Henri Poincaré. What does it actually prove? It proves that any classical system evolving according to determinist laws ends up returning to a state close to its initial state, after a longer or shorter (but never infinite) time. In other words, entropy could diminish and get closer to its initial value. For example, a gas that has expanded would return to its compressed configuration if we waited long enough. This would help to fix a flat tire without a pump, using just a rubber repair patch, some glue, and a great deal of patience. In fact, repetition time is extremely long, greater than the age of the universe as soon as the systems in question contain a few dozen particles. Repetition, as explained by the Poincaré theorem, never has time to occur for systems at our scale; it is the equivalent, for us, of de facto irreversibility.

From there it is a short step to saying that the irreversibility of phenomena is only an illusion proper to our scale of observation. Some physicists who did not accept the idea that irreversibility could come from our human incapacity to grasp all the processes happening in the microscopic world did just this. According to them, something essential escaped physics. Ilya Prigogine, for example, considers that macroscopic irreversibility is the expression of a random nature already happening at the microscopic level: "the statistical description introduces irreversible processes and the growth of entropy, but this description is not due to our ignorance or any anthropocentric feature. It comes directly from the nature of dynamic processes."[3]

Thus, instead of saying that there is no arrow of time, and that the macroscopic level creates the illusion that there is, we can proclaim that there is an arrow of time, and that the microscopic level creates the illusion that there is not. It still remains to determine with precision how the arrow of time could pierce the edifice of microscopic physics, otherwise harmonious, notoriously indifferent to the message of irreversibility it carries.

Other interpretations of the arrow of time rely on quantum physics, which uses mathematical data called the system wave function in order to describe a physical system's state. In general, the system wave function is a sum of several distinct terms, each of which corresponds to a possible value of a physical property of the system (its position, its energy, etc.). One of the troubling originalities of quantum physics is that when we measure an aspect of a system—its energy, for example—a violent change in the wave function occurs: only one term of the sum it holds remains, corresponding to the measured value of the energy. We say that the wave function has been "reduced" by the measure. The choice of the term of the remaining sum after this reduction is totally random, the wave function before the measure allowing only the calculation of the probability that such or such a value is chosen. If the system is measured, only one of the a priori possible results is produced. The mathematical description of the system is then modified, as if the act of measuring it implied the production of an irreversible "mark" on the system.

But recently some researchers have proved that the reduction of the pack of waves is actually due to a mechanism that physics can describe. Their theory, called "decoherence," explains why macroscopic objects behave classically, whereas microscopic objects—atoms and other particles—behave quantically. It says that the "environment," all that surrounds the objects, intervenes—for example, the air surrounding them or, if a vacuum is created, the ambient radiance. Interacting with their environment makes macroscopic objects quickly lose their quantum properties. The environment acts as an observer that permanently measures systems, eliminating all superimposing at the macroscopic level. This process of decoherence has been caught in the act: several recent experiments have enabled the exploration, for the first time, of the transition between quantum and classical behavior.[4] We then start to understand how decoherence can protect the classical character of the macroscopic world. It could also give an explanation of irreversibility in the quantum world, which would look like thermodynamic irreversibility: the evolution of the wave function would actually be reversible, even when being measured, but our macroscopic vision would prevent us from seeing this reversible aspect and would create an apparent irreversibility that would be due, as in thermodynamics, to the impossibility for the observer to know the configuration of a very large number of degrees of freedom. Once again, irreversibility could not be allocated to the physical systems, but would come from the limited description we can muster.

Some cosmologists have suggested that the arrow of time could come from the universe's expansion, which would orient all physical processes according to an irreversible flow.

This might seem contradictory, because general relativity equations are temporally symmetric, but their cosmological solutions, which rule the universe's evolution, are not. The universe they describe is either expanding or shrinking, and is expressed by the existence of a cosmic arrow of time. Some physicists even wonder if this arrow might be the master arrow among all the arrows we have noted.

None of the explanations put forward up to now can be considered complete, however.[5] For the time being, there is no theoretical unity around the problem of the arrow of time, especially because some strangely behaving particles entered the game and complicated it. These particles have a weird name: they are called neutral kaons. They deserve a "little detour," as the tourist guides would say.

16

the kaons gang turns time upside down

This violates the right of being neutral.

VICTOR HUGO

The articles are riddled with mistakes,
Only one person never made any: me!

WOLFGANG PAULI

Physicists grant a lot of importance to the notion of symmetry. We can even defend the idea that twentieth-century theoretical physics has been dominated by this concept, and even more by its twin brother, the *break* in symmetry. Research on symmetries and evidence of small violations has been at the origin of important discoveries, including in particle physics.

Of course, geometric symmetries—of the sphere and the cylinder, for example—come first to mind. But others, more abstract and with wide theoretical significance, are commonly used by particle physicists. Three of them that are deeply linked by a fundamental theorem of physics directly or indirectly apply to the question of time: "time reversal," "parity," and "charge conjugation."

The "time reversal" operation, noted *T*, involves imagining on paper (in equations describing a phenomenon) that time flows from the future toward the past.[1] This is like reversing the direction of the movement of all bodies taking part in this phenomenon, or like showing the film of this event backward. The consequences of such an operation are then examined: if the phenomenon obtained after the time reversal is as physical as the initial phenomenon, it means that the equations in question are reversible vis-à-vis the time variable. For the phenomena they describe, the orientation of the flow of time is arbitrary.

"Parity" is the operation that consists of looking (once again on paper) at the image of a given phenomenon in a mirror. It will be called *P*. Let's use the example of a physical phenomenon involving the collision of particles. Applying operation *P* to this situation means picturing what this phenomenon would become if it were observed in a mirror. The nature of the particles would remain, but their positions would be modified due to the inversion of right and left. The point of all this is to know if the "new" phenomenon can or cannot happen in nature or in a laboratory. If the answer is yes, we say that the experiment respects *P*-symmetry. In the contrary case, we say that it violates or breaks it.

Every particle is associated with an antiparticle that has the same mass and all of whose charges, including the electrical charge, are opposite. The "charge conjugation" is precisely the operation of transforming (still on paper) a particle into its antiparticle and vice versa. For example, it transforms an electron into a positron and a positron into an electron, the proton into an antiproton and the antiproton into a proton. This operation is noted *C*, for "charge," because of the reversal of the signs of the charges between particle and antiparticle.

Let's start with a real experiment of colliding particles. First we carefully record the speeds and the positions of each particle that plays a part in the experiment. We then apply operation *C*: each time we meet a particle, we replace it with its antiparticle and force it to follow the exact same trajectory as the particle, but in reverse. For example, if we look at a collision between a proton and a neutron, operation *C* will describe the exact same collision, except that it will happen between an antiproton and an antineutron. If once the operation is carried out, the new experiment can occur, we will say that in this case too the experiment respects *C*-symmetry. If the opposite is true, we will say, as usual, that it breaks it.

Of course, three operations—*C*, *P*, and *T*—can be combined as often as desired, and in any order: for example, starting with *T*, followed by *P*, and then with *C*. We thus complete the "*CPT* operation," which is very important; it does not modify any of the known laws of physics! In other words, if we project in reverse the film of the image in a mirror of any physical phenomenon in which particles and antiparticles have been exchanged, we observe a phenomenon as likely to happen as the one we started with, and ruled by the same dynamic. This is no accident. In 1940 Wolfgang Pauli was already able to demonstrate that invariance by *CPT* of the dynamic of physical phenomenon must be postulated in any "reasonable" physical theory, because it expresses, in the most formal way, the good old principle of causality! It thus constitutes the core of modern physics. Consequently, if a violation of the *CPT* invariance happened to be observed, the very foundations of the standard model would collapse. But what does this essential invariance mean in practice? Very simply, it means that the physical laws that govern our world are strictly identical to the laws of a world of antimatter observed in a mirror and where time would flow backward. Thus the link that exists between the flow of time and antimatter is confirmed. One of the consequences is that the mass and longevity of the particles have to be strictly equal to that of their antiparticles.

For a long time physicists believed that all laws of physics respected each of the three symmetries, including *P*-symmetry. Isn't it obvious, when we see a layout of objects in a mirror, that we could *also* arrange this layout in reality? Though it was discovered in 1957, to the surprise of particle physicists, that weak nuclear interaction, also responsible for β radioactivity, does not respect the *P*-symmetry![2] In other words, the image in a mirror of a phenomenon ruled by this interaction corresponds to a phenomenon that does not exist in nature, nor can it be produced in a laboratory.

Physicists quickly reassured themselves by demonstrating that the violation of *P* in the process governed by the weak interaction was perfectly compensated by a simultaneous violation of *C*, so that the global *PC* symmetry remained preserved; if these two operations are successively done in any order, we find the same physical laws. The invariance of *CP* combined with the invariance of *CPT* established as a fundamental principle guaranteed *T*'s invariance. But this conclusion lasted only a few years. In 1964 a new surprise came along: an experiment led by James Christenson, Jim Cronin, Val Fitch, and René Turlay enabled the discovery (by chance) that when certain particles (called neutral kaons) decay, they do not totally respect the CP invariance![3] The death of these particles, perfectly natural, is therefore also a "violent" death. But if we assume that the *CPT* invariance is respected in this process as in all others, we have to admit that the *T*-symmetry is also broken, in order to perfectly compensate the *CP*-symmetry. This would mean that there is an asymmetry between the past and the future for neutral kaons, a kind of microscopic arrow of time!

In 1998 the European Organization for Nuclear Research (CERN), the world's largest particle physics center, conducted a gigantic experiment it called CPLEAR (charge parity low-energy antiproton ring) and came to the very same conclusion. It has been known for a long time that, through time, neutral kaons transform into their own antiparticles, which then transform back into neutral kaons, but the CPLEAR experiment was able to show that the speed at which a neutral kaon transformed into its antiparticle was not exactly the same as the opposite process, contrary to what the *T*-symmetry anticipated. That's how the difference between a microscopic process and its temporal opposite had been measured for the first time. But the real reason for this concomitant violation of *T* and *CP* is not really understood.

The question is not resolved, especially since it didn't stop at this point. A new experiment completed in the United States called Babar has just established that other particles, "beautiful mesons," do not respect the *CP*-symmetry when they decay into other, lighter particles.[4] This result, predicted by the standard model, solves a thirty-seven-year-old enigma of physics. Indeed, since the initial observation of the *CP*-symmetry violation in the decay of neutral kaons, physicists were wondering if this phenomenon was relative to them or if, on the contrary, it was a more general rule.

This result could also help solve an old physics problem that is so fascinating that it must be mentioned here even though it isn't directly linked to time. We know today that the universe is almost exclusively composed of matter, but this hasn't always been the case: in the distant past, the universe contained nearly as many antiparticles as particles. The question—in the form of an enigma—is the following: knowing that particles and antiparticles have symmetrical properties, how is it that our world is composed of the first rather than the second?

To better understand, let's retrace the facts of the case. Galaxies are islands of matter in space. Is it possible that some of them are exclusively composed of antimatter? This hypothesis did not stand up to observation, because the existence of collisions between matter galaxies and antimatter galaxies should produce, by annihilation, very intense radiation spreading in all directions through space, something that has never been observed. Besides, nobody has imagined a process that could have completely separated matter from antimatter in such a way that they could afterward form large homogeneous structures. We are thus doomed to admit the existence of a radical dissymmetry in our universe: matter is dominant, antimatter has been eliminated from it.

The standard model of cosmology predicts that the primordial universe should contain equal amounts of matter and antimatter, annihilating and creating each other in a cloud of photon gas. The expansion of the universe has progressively cooled off this environment, diminishing the available energy in a given volume. The largest particles, which demand more energy to materialize, disappeared first, giving birth, through their dematerialization, to other, lighter particles. The lightest have survived, their mutual distance progressively increasing due to expansion. Their density decreased in correlation, making the annihilations less and less frequent. But all this hasn't been enough to break the balance between the quantities of matter and antimatter. We just have to imagine a mechanism in the universe's distant past by which the second could have disappeared to the benefit of the first.

In 1967 Andreï Sakharov was the first physicist to consider the possibility of a slight surplus of matter over antimatter, also indicating the three necessary conditions for the apparition of such a dysymmetry.[5] Among them is none other than the violation of the *CP*-symmetry, hence the violation of the *T*-symmetry! The structure of the macrocosm was thus linked to the laws of the microcosm. Sakharov explained that if these three conditions are met, the number of protons and neutrons produced at the birth of the universe (which constitute matter today) might have been largely superior to the antiprotons and the antineutrons. After matter annihilated antimatter, all the antimatter would have disappeared, but the surplus of matter, which was extremely weak (a proportion of about 1:1 billion), would have survived: it would constitute the matter that we observe today as well as the matter we are made of. The current universe's matter would thus be the improbable survivor of a mass slaughter. This elegant hypothesis has been partly confirmed by the experiments that were discussed earlier in this chapter.

And that is how, by first investigating the violation of minuscule objects, we are led to wonder about the overall structuring of the primordial universe, fifteen billion years ago. On the one hand, this shows that the infinitely small and the infinitely large are interdependent, linked by the same bonds, and on the other hand, that some phenomena we can observe today reveal traces of the past in the universe. Would physics, sparing in its privileges, forbid time travel except for itself?

17

2002: does cosmic time accelerate?

It is very simple; you just have to
push the pedal to the metal.

LOUIS LACHENAL

—I don't know what destiny is.
—I am going to tell you. It is simply the accelerated
form of time. It is awful.

JEAN GIRAUDOUX
LA GUERRE DE TROIE N'AURA PAS LIEU

WE NO LONGER have time to take our time. The logic of the chronometer has invaded chronology, transforming us into slaves of speed, into "turboninnies" or "cyber silly geese" to speak like the late Gilles Châtelet.[1] Modern life indeed imposes an unprecedented rhythm on us: a rhythm of constrained, compressed, framed time that gets transposed into an ensemble of subjections and imposed figures. Then, as if out of breath, we exclaim, "Time has sped up!"

At the same time, some astrophysicists (who are not atrocious physicists) have discovered that the expansion of the universe speeds up too! Would everything therefore go quicker? Let's take a slower, closer look at things.

Thanks to new means of detection, numerous data has been gathered and astrophysicists have managed to carefully analyze light emitted from very distant supernovae referred to as Type Ia. What they have discovered still amazes them. The Type Ia supernovae are linked with exceptionally brilliant explosions. They are made up of small, very dense stars called white dwarves, and more massive companion stars, most often red giants. The white dwarves have nearly the sun's mass, but are concentrated in a volume the size of the earth's, making their gravitational field very intense. Hence they possess a terrible voracity; they tear off and gluttonously absorb their companion's matter. This orgy increases their mass and their density, until it provokes a gigantic nuclear explosion. We can see this because they emit a very bright light, called a Type Ia supernova. Such an object can shine for several days with a light as bright as a billion suns.

Astrophysicists are interested in these events because of the fact that such occurrences are used as luminous standards: they are nothing short of the candle standard that enables scientists to survey the universe on a large scale and measure some cosmological parameters. This virtue is due to how their "light curves" look very similar, appearing first as a brilliant peak that lasts several weeks, followed by a slow dimming. This resemblance is easy to explain: the different light curves come from similar objects whose explosion mechanisms are identical, so much so that they produce the same temporal structure. Therefore, the single difference between two light curves can come only from a distance: the farther away the supernova is, the weaker the light we receive from it is. Measuring the intensity of this light, we can then calculate the distance of the star that emitted it.

More than fifty of these Type Ia supernovae have been studied, with distances of up to six or seven billion light-years away.[2] The results are surprising: these supernovae seem to be even farther away than expected! To be more precise, their position lets us assume that the universe has been in an accelerated stage of expansion for several billion years. What does this mean? In the process of expanding, gravitation, which is always attractive, acts as a brake: it tends to bring massive objects closer to one another. But what these new measurements seem to show is that another process is opposing gravity by acting as an accelerator. Everything happens as if a kind of "antigravity" had taken control. This new information has to be taken seriously, because in April 2002 another team of researchers found similar results using another method.[3]

These conclusions, if they are confirmed, oblige us to attribute a positive value to the "cosmological constant," the parameter that Einstein introduced into the general relativity equations out of desperation. At the time he wrote them, the universe's expansion was not yet discovered, and he, like nearly all of his colleagues, was convinced that the universe could only be static.[4] Yet the universe can be stationary only if gravity, by which matter attracts matter, is compensated by something else. Otherwise it ineluctably collapses upon itself. Einstein thus introduced a new term into his equations, the cosmological constant, whose influence is similar to an energy that is exercising negative pressure; it corresponds in a way to a repulsive gravity, or more exactly, a repulsion of space toward itself.

After Erwin Hubble had discovered drifting galaxies and the expansion of the universe, Einstein sided (after much hesitation) with the idea that this constant had lost its reason to exist and that he had been wrong to introduce it into his own equations. He published a famous article on the subject, an article that is often quoted but seldom read in its entirety.[5] But most cosmologists today think that the cosmological constant has no reason to be zero. It could have some a priori value.

When it is positive, the cosmological constant corresponds to a kind of repulsion of space toward itself. It then should transmit a kind of acceleration of the expansion of the universe. But things are not so simple. We should guard against granting the cosmological constant virtues it might not have, because we cannot exclude the existence of still unknown "substances" whose effects would be similar. For example, some "exotic" matter representing up to 70 percent of the total contribution of the universe's mass might also be the agent of acceleration for the expansion of the universe on the condition that its physical properties would not be like standard matter (in particular, it should have negative pressure, that is to say, apply a repulsive force). The question of the physical nature of this new matter is posed. One of the candidates is the quantum void (which is not at all nothingness), but nothing allows us to claim that it has a gravitational influence on the universe. Other candidates are considered—a scalar field, for example, which is sometimes called the "quintessence," whose structure would be such that it could be at the origin of this accelerated expansion. Whatever the case may be, it now seems to be accepted that ordinary matter, matter that is blessedly composed of atoms, is only a fringe of the universe's matter—its visible form, so to speak.

Does this acceleration of the universe's expansion affect the flow of time? According to some theories currently sketched out, such as "quantum cosmology," the universe's expansion could be the real engine of time:[6] if it accelerates, if the engine of time is gearing up, the flow of time should also "accelerate" (if we are even able to give a meaning to this expression).[7]

Could this phenomenon of the acceleration of cosmic expansion be at the origin of the impression (unanimously shared, apparently) that our lives are getting quicker? Of course not, because it is imperceptible at our scale and remains to be confirmed. Under the pretext that we do everything more quickly, that everything accelerates around us, we ceaselessly claim that time itself is accelerating. Time, though, does not accelerate. It is what it is, indifferent to our restlessness, independent from our acts, our moods, our impatience. An hour lasts an hour, whether we pass it lawn bowling, suffering hell in the dentist's hands, or dancing in the arms of our beloved. Why should the flow of time depend on our schedule? For no reason, of course. But it is always the same; we always tend to grant time with the characteristics of the phenomena it contains. Yet it is not because the amount of time we need to do such and such a thing decreases that time itself is going quicker.

We think we see an acceleration of time where there is only a fragmentation of reality, a growth in production, an exuberance of the becoming, a "second coming" of telecommunications: more and more goods, less and less work time, the appearance and disappearance of things with great frequency, the apparent elimination of distances. This insistence on confusing time with speed—or time with agitation—reveals much about our relationship to modernity. If we identify time with the materiality of change, the dynamism of our actions, the rhythm of our exchanges, isn't it because we believe that the more innovations there are, the more reality multiplies and diversifies, the more temporality in action there is?

The explosive trend of our societies thus seems to lead to confusion between time and what we produce in it. Although five hundred million years separated the invention of fire from the firearm, a mere six hundred years was enough to go from the firearm to nuclear fire. These days, producers propose a "new generation" of products every year. What is "out-of-date" increases so quickly that soon the speed of light, Hershey bars, and especially the Rolling Stones will remain our only standards of invariance.

We cannot help it—we are fascinated by the idea of speed. There is certainly a hidden metaphysics behind that; because we have the feeling to be secretly separated from ourselves due to our own expectation, we also have the feeling that what would shorten this expectation would bring us closer to ourselves. That is why each time we deal with acceleration and quickness, we are like sailors feeling the wind kick up, already scanning the horizon, as if the promised land, the end of the waiting, were right there.

18

some time. . . . only from time to time?

At this banquet, Mister Ulysses
Had a serious case of hiccups,
And since it is very rude
To hiccup in a party,
He drank to God's health,
Hiccupped, and bade farewell.

MARIVAUX

See the mountain
Don't see the mountain
See the mountain again.

QING-DENG

PHYSICS, WHICH IS not keen on sterile complications nor useless hypotheses, preferred considering, throughout its history, that space and time are "smooth" entities and that we can thus represent them with continuous magnitudes. There should be space everywhere and the past forever, holes being impossible, so that we can consider lengths or durations as brief as we want, without ever reaching any limit.[1]

This idea seems so natural that it slipped into common culture like a letter into a mailbox. Wouldn't renouncing it create enormous difficulties? Let's assume that time is discontinuous—"discrete," as physicists would term it—that is to say, it is constituted of particular instants, separated from one another by durations deprived of time. How could the flow of time ceaselessly stop and then ceaselessly start again as if it were having the hiccups? And how long would the periods deprived of time last? It seems impossible to conceive that time would exist only occasionally! The idea of a discontinuous or intermittent time, with *ticks* and *tocks* but nothing "in between," brings us inevitably to the difficulties of conceiving of stopped time, whether the stop is transitory or definitive.

But will this idea be rejected endlessly? After all, the history of physics already documents cases in which discontinuity occurred where we only pictured continuity. Consider the discovery of *quanta* at the beginning of the twentieth century. Until then, physicists were convinced, with every apparent reason, that the continuity of space and time triggered the continuity of speed, and from there, the continuity of energy. But Max Planck's research on the black body suddenly convinced them that they had been wrong for a very long time: exchanges of energy between radiation and matter can happen only in discontinuous packs. There is thus no reason to be too scrupulous a priori when evoking the question of continuity or discontinuity of space and time. We are at least authorized to pose the question without feeling we are committing a crime of *lèse-majesté*.

This is even more so since the supposed continuity of space is creating some torments, because it enables us to take into account minute, even nonexistent, distances. Let's consider, for example, the electric field produced by an electrical charge, say an electron, at a certain distance from the field: this field, varying like the inverse square of the distance, becomes infinite when it is cancelled out. Such divergences or "singularities" lead to mathematical difficulties that we usually eliminate with the help of different processes of calculation that either abolish or neutralize them.

But we can also consider another, much more audacious angle. It comes down to imagining that space itself could be discrete, noncontinuous—that is to say, structured in a kind of network, whose weave, thin but not nonexistent, would represent a minimal distance under which it would be impossible to decrease. Any divergence would then be avoided. But new problems immediately emerge. For example, such a network would introduce privileged directions that would destroy the isotropy of space, its invariance by rotation. Yet this invariance, along with other symmetries of the same type, plays a fundamental part in physics by imposing very restrictive rules of conservation.[2] The discontinuity of space theory thus appears to be a dead end.

But the situation has recently and suddenly changed; a new angle has appeared in the extensive work completed in the 1980s by Alain Connes. This concerns the so-called noncommutative geometries, which enable us to consider structures that present a discontinuous character without breaking fundamental symmetries. To build these new geometries, we have to replace the usual spatial coordinates, which are ordinary numbers, with "algebraic operators," which have the property of not commuting between one another. This means that the order of their application matters. These algebraic operators are specific: they verify some relations that define the properties of space at a very small scale. There is thus always some "space," more precisely a spatial structure, but when closely examined, this structure does not have its ordinary properties. The beauty—and the theoretical strength—of these new constructions is that at a larger scale they restore the usual properties of space. They then invite us to consider space as we know it actually emerging from an underlying structure very different from itself. For example, the smooth aspect of space, its apparent continuity, would be seen as foam floating above a discontinuous network of points. Everything happens as if the universe first had to "blow up," before space could assume the calm aspect we know today. We can compare this situation to what happens when you look very closely at a TV screen: with your nose glued to it, you see only dots of three different colors, but no real picture; the picture progressively appears as you step back, and with it new colors. In the same way, space could appear with its continuous characteristics only after the universe had reached a certain size.

With time and space being linked together, could such ideas apply to time? Could time too have been discontinuous at a very small scale? Could elementary durations take only certain particular values? It is better not to jump to conclusions on this subject, out of common sense. After all, it is not impossible that some equations are much smarter than we are, or not yet intelligible, that they drill escape routes where our prejudices are blocking us, that they express situations we are not yet able to think of. An equation is sometimes much more than an equation.

Thus, because we are faced with calculations that deny our experience of time and whose conclusions are at first absurd—time is passing and then is not passing anymore, then it decides to start passing again, and so on—we should temper our spontaneous skepticism.

Being open-minded is also necessary to accept some recent theories that show not only one time, but several times—at the same time. Might the universe be dancing the two-step? Or even three?

19

dance of the superstrings and the several-steps waltz

He was a faithful man.
The problem was that he had too many wives.
HÉLÈNE WEIGEL
(BERTOLT BRECHT'S WIFE)

Up until now we have considered that physics, being faithful to common experience, assumed that space had three dimensions and that there was only one time. Of course time is relative, of course it is coupled with matter, but it is definitely unique. However, if we examine some very daring theories being developed in the field of particle physics, we discover that this situation could change one day, despite enormous reticence. Could there really be several times at the same time?

Particle physicists are interested in objects, in particles, that cannot be seen because of their extremely small size. They are also interested in their mutual interactions. Four of them are really fundamental: gravitation, electromagnetic interaction, and two nuclear interactions that function only on a microscopic level. The "weak" interaction governs some radioactive processes; the "strong" interaction binds the components of the atomic nucleus.

In the 1980s a tremendous discovery was made: it was proved—first theoretically, then experimentally—that the electromagnetic interaction and the weak nuclear interaction, though apparently very different, were not independent from each other. At one point in time—long, long ago—they were one, the "electroweak" interaction. This important discovery illustrates the outcome of a very clever use of the symmetry concept: we have seen that the structure of the interaction between particles can be deduced from their properties of symmetry.[1] The preliminary identification of the symmetries associated with electromagnetic and weak nuclear interactions later enabled them to be unified under a theoretical point of view, by putting them in the same "mold." This fertile procedure has been extended to strong nuclear interaction. The result constitutes the current "standard model" of particle physics, which has been carefully tested by the LEP (large electron positron ring), the CERN's large collider. Because of this success, we can claim that forces are not ingredients that have to be arbitrarily introduced into theories next to the particles they govern, but rather that they are a result of properties of symmetry to which these particles abide.

Thanks to the standard model, physicists have managed to describe the behavior of particles at a level of distances on the order of 10^{-18} meters.[2] But at much smaller distances, the equations no longer work: a new physics appears to be necessary; its elaboration will imperatively have to take into account gravitation, which has been put aside up until now. This "widening" of physics will be possible only if we modify our representation of fundamental objects, as well as space and time.

An apparently very promising lead is now being studied—the superstring theory. Its foundations were elaborated in the 1970s, with the objective of building a general frame that would be able to encompass quantum physics, which describes elementary particles, and general relativity, which describes gravitation.[3] These two theories actually seem conceptually incompatible: quantum particles are described in a flat space-time, which is absolute and rigid, while the space-time of general relativity is supple and dynamic. In the superstring theory that transcends both of them, particles are no longer represented by objects with zero dimension, but by objects that are slender and without thickness—superstrings—that vibrate in different space-times having more than four dimensions. More precisely, the theory replaces all punctual particles we know of with one stretched object, the superstring. This superstring can be open (meaning it ends at two extremities) or closed on itself,[4] and its different modes of vibration correspond to different possible particles: one mode corresponds to the electron, another to the neutrino, a third to the quark, and so on. The usual particles correspond to modes with the lowest frequencies. Other, heavier particles correspond to modes with higher frequencies. They remain to be discovered.

To understand how the (crazy?) idea to increase the number of dimensions of space-time has sprouted, we have to return to the brilliant 1920s. Einstein was wondering if electromagnetic effects could be looked upon as a geometric property of space-time. Such an idea had worked pretty well for gravitation, which Einstein himself had geometrized through his general relativity. Yet electromagnetism and gravitation have some similarities, at least because their force varies inversely to the square of the distance.

Determined to unify them, in the early 1920s Theodor Kaluza and Oscar Klein proposed a revolutionary theory in which electromagnetism and gravitation were put in correspondance.[5] They noticed that writing the general relativity equations in a space-time with five dimensions (four of space, one of time) enabled them to obtain, after projections on more restricted space-times, the usual equations of general relativity, and an additional equation equivalent to Maxwell's equations. Some kind of unique force in a space-time with five dimensions thus appeared equivalent to two interactions (gravitation and electromagnetism) in a four-dimensional space-time. Hence the idea was born that the unification of the interactions could necessitate an "enrichment" of the topology of space-time; it would then be up to its promoters to explain why we do not find the additional spatial dimensions. Kaluza and Klein suggested that the fifth dimension of their theory was rolled up on itself on a minute scale and that it was therefore imperceptible, in the same way that a piece of fabric (a three-dimensional object) appears to us like a two-dimensional object because of the extreme relative thinness of its threads. Space-time could thus appear to lose dimensions that nonetheless exist on an ultramicroscopic level.[6]

The superstring theory, which aims at a coherent description of gravitation using particle physics, recycles Kaluza and Klein's theory but postulates a ten-dimensional space-time, some dimensions being "compactified," that is, folded on themselves, so that they would be imperceptible on a human scale.[7] These additional dimensions enable us to eliminate the infinite quantities that calculations bring about when concerned with the interactions happening on very small spatial scales. Actually, the conceptual frame in which superstrings are described is not imposed in just one way. Several possibilities coexist, but they all require the existence of gravitational force, as described by Einstein's theory of general relativity. In other words, in superstring theory, gravitation, instead of being simply stated, acquires the status of a prediction drawn from the very principles of the theory, and that is of course what gives it its formal beauty, its "magic," as certain physicists even say.

But we should not forget that some experiments are essential to validate such a beautiful construction. How can we demonstrate some new physical phenomena linked to the existence of additional dimensions of space-time? When the superstring theory was conceived, physicists pictured that the size of the additional dimensions could only be the smallest possible length in physics, which is the Planck length, close to 10^{-35} meters.[8] In such conditions, any manifestation of a physical phenomenon in one of these dimensions would appear totally beyond the means of current observation, including the most powerful particle accelerator. The LHC (large hadron collider), which will start functioning at the CERN in Geneva in 2007, will probe distances at the scale of "only" 10^{-19} meters by provoking collisions between two beams of protons of 7 TeV each.[9] Such distances, ten millions of billions of times longer than the Planck length, are still too significant for us to see the slightest effect linked to the existence of the superstrings. At least that is what we thought for a very long time.

In 1996, lo and behold, Edward Witten showed that the size of the superstring is in fact a free parameter of the theory and that there is therefore no reason a priori to peg it to the Planck length.[10] Since then, many theoreticians are getting passionate about the idea that the additional dimensions of the superstring theory could be around 10^{-18} meters. If they are right, some of the effects linked to the superstrings could be detected thanks to the LHC.[11]

The existence of additional dimensions, all believed to be spatial, makes us think that all this should not have an effect on the question of time. But because we could imagine that, one of the additional dimensions would be at least temporal and not spatial, which would mean that time has several dimensions, and that only one, corresponding to physical time, would not be rolled up on itself. This idea is not given much credence, because it would demand a disruption, a revision of our way of thinking. How could we actually understand the existence of several coexisting times? This question becomes even more disconcerting if we consider rolled-up temporal dimensions. Creating loops, their structure would break causality, forcing particles to periodically return to their past. They would thus be authentic time-traveling machines, if not for us, at least for some ultramicroscopic objects.[12] Accepting them would come down to renouncing causality as we understand it today.

But here again we should be careful not to conclude anything and, above all, not to snigger, because research on the subject is rapidly evolving. Today, theoreticians would like to stop hampering theory by forcing it to operate in a given, a priori, space-time. They would rather try to enable it to create its own spatiotemporal arena starting from a configuration deprived of time and space, such as those suggested by the noncommutative theories we just evoked. They might then manage to show that the space of small birds and the time of the clock are just convenient notions that come from a structure that does not contain them at a very small scale, a kind of "soup" of moving superstrings. They both would actually be products of the theory and not theoretically presupposed entities.

If the equations continue to be so daring, time could soon stop being what it is—at least in the complicated calculations of physicists. Nobody knows if this new understanding will end up affecting daily life. In order to be on time, will we have to wear a watch on each wrist?

20

theories seeking origin of time, desperately

I will give you some damn Nordic mist.

MARCEL AYMÉ

Please know that Rose Sélavy's famous act
is recorded in the celestial algebra.

ROBERT DESNOS

SCIENTISTS FACE PROLIFIC difficulties with the general notion of origin, whatever the subject—difficulties of matter, life, consciousness, man, and thought—because science, in order to grow, needs a reality, something "already there."[1] But origin, precisely, is not part of the "already there." It corresponds to the emergence of a thing in the absence of that thing. It is thus a meeting point between being and nothingness, a contact between all and nothing: there is nothing yet, and something appears. The origin is a nothingness out of which something has to come, as if being were already contained in it. It is thus an ontological singularity because of the fact that it assumes nonpresence in putting in presence and, at the same time, the potentiality of presence in the heart of absence. Yet science is not able to grasp this singularity, and even has a hard time giving it a status that enables it to be treated. As soon as science speaks about it, it implicitly invokes a kind of "Jupiter's thigh" made up of prior ingredients that have to be added to the story to understand the origin in question. To science every beginning looks like a consequence: it ends something. So would time be an exception?

Most physicists now agree on particular models of the universe, based on the big bang, in which reigns a "cosmological" time linked to the expansion of the universe and to which general relativity gives a status. This cosmological time shares the property of being universal with Newtonian time: observers who are not submitted to any acceleration or any mutual gravitational effect can actually synchronize their watches and have them remain in phase all along the cosmic evolution. Thanks to this time, we can describe the major phases of the history of the universe that, according to astrophysicists, unfolded over fifteen billion years: (1) matter eliminates antimatter, its antagonistic double; (2) then light splits from matter, making the universe transparent to its own light and matter free to restructure itself; (3) then the galaxies are born, the stars and all the shapes that fill the night sky. That's how some genealogies, some genetic links are declined: stars are the mothers of atoms, their ancestors are clouds of dust, whose matter comes from the primordial universe.

The universe, it is now believed, had a history. So does this mean it had a beginning? As soon as we evoke the idea of a beginning, the question of origin arises. It immediately overwhelms us. We are incapable of knowing if the material universe had a temporal prelude: did it appear in a time that preceded it, or was its emergence simultaneous with time's emergence? If we suppose that a time existed before the universe, what prevents us from claiming that this time was already in itself a universe? That is what Kant noted in his *Prolegomena*: "Let's admit that the world has a beginning: since this beginning is an existence preceded by a time where the thing is not being, there must have been a time where the world was not, that is to say, an empty time. Yet in an empty time, it is impossible for something to be born."[2] Speaking about the beginning of time creates an aporia: it comes down to situating time within time. Only myths seem able to transcend this contradiction.

In a more general way, questions about origin quickly turn into a chicken-or-egg story that forgets to mention the rooster. Most often, these questions drag along a host of big words, struggling for precedence and holding one another by the chin: *creation, emergence, purpose, chance, necessity,* and even *God* in extreme situations.

What is the real origin of cosmological time? How did it start? The answers to these questions seem out of reach. Some astrophysicists, however, regularly promise that they soon will be able to find answers. We are on the verge, they say, of putting the universe at equations' length; the unveiling of the scenario is about to be completed. This phraseology is not new. The power of physics has always aroused enthusiasm far beyond what its theories can offer. Contemporary physics does not escape this rule.

Sometimes, blinded by success, it finds itself exposed to the risks that go with victories. Ready to announce its future success, it is inspired by "satisfied" thinking: the always beneficial kind that pushes it to make daring hypotheses, or the more toxic kind that lulls it into the arrogant certainty of reaching its objective.

So what exactly do we know about the "origin of time," if the expression means anything? Like that of the universe, it gets lost in the auroral mists of a primordial universe. Neither general relativity, nor quantum physics, nor an eventual synthesis of the two can currently enable us to describe the apparition of the universe as a physical event. Language is also powerless to say anything on the subject: it is impossible to say with words something that would resemble "the history of the birth of time," since the universe is, among other things, time, and nobody sees how we could speak about the creation of time without time.

But here again I can hear some of you saying, "Aren't the equations of traditional cosmology able to go from the present back to a 'zero instant'?" So be it. This "zero instant" can be called "origin" if we really want, but without forgetting that it actually corresponds to a situation where equations cease to be valid. In other words, this first moment is not really a moment, in the sense that it does not correspond to any actual moment of the universe's past.

Contemporary physics is undoubtedly able to describe the universe backward. But when we extrapolate its laws onto the past, we end up finding a state of the universe in which physical laws struggle with one another because of the incompatibility of the principles of quantum physics and general relativity. These two theories collide precisely because of problems dealing with space and time. Each time they attempt some vague nuptials, certain singularities (that is, infinite divergences) spontaneously appear: the space-time they produce becomes an impossible sea to sail on.[3] This situation authorizes all conjectures without saying how to determine which one seems to be the most in accordance with what was the universe's most ancient past.[4] We therefore do not know anything—today—about the origins of the universe, nor about the origin of time, whether the term "origin" is understood in a chronological sense or in an explanatory one.

It is just as perilous to wonder what could have been there before the big bang.[5] Of course there are some theories, including the superstring theory that we just evoked, which enable us to consider a "pre-time" different from the usual physical time, but this notion, far from answering the question, only shifts it: what was there before this famous "pre-time"? And even before? If future laws of physics—a potential "everything theory"—enabled us one day to describe the origin of time, we would immediately ask: what is at the origin of these laws? And at the origin of the origin of these laws? Any beginning, far from being a foundation, always demands to be founded in itself, somehow like a regression of the conditioned to its condition. Any progress in this ontological swamp forces us to invoke, at each step along the way, a new gift from the Gods: quantum emptiness, the explosion of a primordial black hole, a collision between two multidimensional superstrings, and so forth. Is the God of the sky and the elements perhaps a caterpillar?

To wonder what existed before time is actually like wondering what is north of the North Pole. In both cases, we can only answer "nothing." By definition, there isn't any period before time, so that the question of knowing what could have happened there is meaningless, just like saying that if there is nothing north of the North Pole, it is because the area we are imagining does not exist or that the words we have at our disposal—including the word "exist"—cannot express it.

Does the question of the origin of time or the universe lead us down the wrong path? Since we are accustomed to thinking in terms of cause and effect, we naturally look for a chain of causality that reaches back in time, a chain that either has no beginning, or reaches a primary cause or a primordial engine (God, for example). Yet contemporary cosmology invites us to consider a universe without any prior cause, in the usual sense, not because this cause would be abnormal or supernatural, but simply because it does not foresee any anterior period in which it could operate. Our minds, smitten with a love for logic, have a difficult time accepting that the meaning of the question they ask could vanish in a vertiginous mess.

If physical time, which we have been investigating up until now, is consubstantial to the universe, it seems that a new time appeared with the human being, a human time, a "time of consciousness" that expresses the ways human beings live and experience life. Does this other time derive from physical time, or does it have an autonomous existence? Before tackling this question, let's linger a little on the various ways we perceive and experience time.

21

chronoclastic spirit, useful watch

It is a weird stroke of fate: every man whose skull we opened had a brain inside!

LUDWIG WITTGENSTEIN

Did I really let out the watch and wind the cat?

GROUCHO MARX

ANSWERING A QUESTION from Bergson, Einstein explained one fine day in 1922: "There is not a time of philosophers; there is a psychological time different from the time of physicists."[1] It really seems that there is a time of consciousness that is radically different from the time indicated on clocks. This "psychological" or "subjective" time would be, according to a now well-established vulgate, a kind of a second time developing alongside physical time. To grasp its substance, the experience of boredom is no longer necessary, but Paul Valery's recommendation is right on target: "Wait for hunger. Prevent yourself from eating, and you will see time."[2] One need not fast for forty days to realize there is a certain truth in this.

The story seems set: a psychological time definitely exists, which is not to be mixed with physical time. The most obvious distinction between these two times deals with their fluidity. Physical time flows in a uniform way, whereas psychological time changes in a rhythmical way; according to the circumstances, it can give the impression of stagnating or, on the contrary, speeding up. If we wear a wristwatch, it is because our estimation of duration is not reliable: we regularly have to set our records straight.

Numerous factors combine to endlessly modify the texture of our psychological time: age, of course, but also our state of impatience,[3] or else the intensity and the meaning the events we are experiencing have *for us*. To try to better grasp them, radical experiments have been implemented (several decades before "Big Brother," but without any camera). For example, "Speleonautes," men and women who chose to live several months in caves or bunkers without a watch or clock, were left only to their biological rhythms. We rapidly learned that their estimation of duration in these conditions deviated considerably from what clocks indicated.

The same phenomenon is observed when the confinement is not voluntary. In his book *Psychologie du temps*, Paul Fraisse recalls what happened in 1906 with the big mining catastrophe in Courrières: after the cave-in, some miners were stuck in a tunnel from which they escaped only after three weeks of rescue efforts. Once freed, they spontaneously declared they felt they had spent only four or five days at the bottom of the mine. Durations, even when lived under duress, can thus be estimated to be five times shorter than what they really are.[4]

Our inability to precisely quantify durations when all outer landmarks have disappeared suggests that psychological time is not only elastic but is also very different in substance from physical time, which, always unrolling identical to itself, has the shape of a filament. Subjective time seems to unfold following a broken line—play the accordion, mix different rhythms, suffer discontinuities. We can then assimilate subjective time to a uniform fourth dimension that would simply be added to space. Its structure looks more like a plaited cord with a very irregular pattern, light-years away from the traditional picture of physical time.

Physical time and psychological time would also be different from each other in their respective ways of "presenting the present." Physical time's present has no duration. It is concentrated on one point, the present instant to be exact, which separates two infinites from each other: the infinity of the past and the infinity of the future. Psychological time mixes a little of the recent past and a little of the imminent future in the very heart of the present. It then creates a certain duration by unifying what physical time keeps separating, temporarily withholding what it carries away, encompassing what it excludes. In physical time, two successive instants never exist together, but when we listen to a song, we clearly perceive that the preceding note is seemingly held captive by the current note, which projects itself into the upcoming note. In our consciousness, the present is touched by the remnants of the preceding instant and the anticipation of the following instant. That's how the brain organizes a kind of continuity linking the immediate past with the present and the impending future (an alliance without which we could not speak of "melody" in music).

Does this mean that the elasticity of our estimation of duration could be explained with purely cerebral mechanisms? The story is more complicated than that: neuroscience specialists have been able to show that this operation requires several cerebral areas, such as the cerebellum and the frontal cortex, but without managing to explain the mechanisms supporting it.[5] It simply seems that the information corresponding to time passing by is not recorded or coded as such. There would not be a "sense of time" comparable to other senses, such as vision, even if some laws relative to our perception of duration have been established.[6] More precisely, our internal chronometer, if it really exists, would not permanently be solicited by consciousness. Indeed, the mechanisms used by our brain to estimate durations seem to get activated only when we are put in a specific situation of waiting—for example, when we are warned that we will have to estimate the duration of a song or of a luminous signal.

To correctly estimate a length of duration, we first have to *concentrate* on this operation; that is to say, we must clear our mind of anything that could distract or perturb it. But this is not enough to become really precise: even when concentrated we are able to perceive only changes and events that happen in time and not "pure" duration. Indeed, when we want to grasp duration for what it is, "it is always the same failure," as Gaston Bachelard explained. "Duration does not limit itself to lasting, it lives! No matter how small the fragment under consideration is, a microscopic examination is enough to read a multiplicity of events; always embroidery, never the real stuff; always shadows and reflections on the river's moving mirror, never the clear flow."[7] Any duration appears to be soaked with the events it contains, as if between the flow of time and us there were always problems on the "line"—crackling, interference, and other sounds.

These line problems are what give psychological time such diverse and complex modulations that nobody ever managed to show how we could "divert" them away from how our brain apprehends physical time. Does this condemn us to admit that physical and psychological times constitute two distinct realities? Maybe. But isn't reaching this conclusion once again jumping the gun? The fact that our way of perceiving durations is filled with psychology, sometimes to the point of saturation, in no way implies the existence of an autonomous psychological time that would flow with elasticity beside physical time. We can also defend the idea that there is only one time, *physical time,* but that we feel obliged to call for another time, *psychological time,* because of our helplessness in unfolding Möbius's strange ribbon that binds, at the seam of matter and spirit, the first one to our subjective perception of it.[8] The infinite variety of our moods, of our states of mind, would disguise physical time until distorting it into a second kind of time.

To get to the point, the existence of a psychology of time is not enough to prove psychological time. It thus seems to be more prudent to suggest that what we call psychological time is only the manifestation of our subjective relationship to physical time.

It is no longer possible to have an overidealistic conception of time, like Kant, who subordinated time to the subject. Is time the simple "subjective condition of our intuition" in which sensations are organized? This definition does not say much about the feeling, essential in each of us, that we are submitted to time as to an *external* power that leads us along. There is thus something that if not enigmatic is at least curious in Kant's conception of time. As Pierre Boutang underscored, "the ambiguity of time in Kant, between his original *a priori* which confines it to a singular mode of human sensibility, and its coextension with all phenomena of the world, has not been explained—or even made clear—by any of the philosophies or cosmogonies in the two centuries after his death." To paraphrase Kant, I am time, and I am also *in* time. But how can we simultaneously think of time as a mode of human sensibility and as a data of the world?

22

endless unfurling of the present instant

If we live in lightning, it is the heart of eternity.

RENÉ CHAR

Were it not for the point, the immobile point,
There would be no dance,
And all there is is the dance.

T. S. ELIOT

In the eyes of a physicist, our consciousness never stops achieving the impossible, since it makes bits of the past and the future, which physical time has never united, coexist with the present moment. The flow of time as we perceive it can only be thought of by invoking this strange intermingling in the very heart of consciousness of sequentially separated elements appearing to be interdependent.

Maurice Merleau-Ponty said, "Consciousness unfolds or builds time."[1] This is what Saint Augustine had already proclaimed. He believed in a "present of the future," which he called *expectation;* a "present of the past," which he called *memory;* and a "present of the present," which he called *attention.* This formulation achieved the feat of making the three "ekstases" of time (to cite Heidegger) communicate in a noncontradictory way. It also managed to express the human experience of time in such a remarkable way that for a century all the different phenomenological schools appropriated and dissected it.

It is certainly because of this continuous connection established in the consciousness between past, present, and future that we have such a hard time directly experiencing physical time, made of punctual instants, lacking thickness. It also explains why we do not feel the searing intensity of the present instant. Indeed the present, as we perceive it, never has the cutting edge of pure brilliance. In general, it reveals itself to us through a representation that wears out its essential vigor: we never perceive moments as singular entities, we do not feel those temporal atoms "with no extended duration" that Saint Augustine spoke of.[2] Everything happens as if our awareness, while being attentive to the world, activated a certain inattention coefficient toward life in order to erase part of the present's radiance by mixing it with what comes before and what comes after. The present is thus distributed on both sides of the punctual instant that constitutes its center.[3] It breaks into two parts, which have the precise characteristic of not being present. The first one is made out of what has just been and is passing by. The second is either a momentum that triggers the future to happen, or a passive expectation of what is going to appear, and often a mix of the two. The present is thus usually fed by an illegitimate assemblage of tension and retention, which smoothes down its potentially explosive capacity.

But there are some situations that do not follow this rule, some of which are tiresome, others enjoyable. Let's start with the tiresome ones, which deal with suffering, including physical suffering. When such a situation is intense, it is expressed as an impossibility of getting detached from the present instant. It exposes the being, strips it, reduces it. There is the unbearable absence of any refuge from time in suffering. We are "glued" to ourselves, incapable of escaping, going forward or backward, or of taking a break. All the acuteness of suffering is actually in this impossible step back: the present imposes itself without allowing any possible distance.

It also happens that the present reveals itself in an ecstatic way, without getting mixed with what precedes or follows it. Who has never experienced these magical moments that Kierkegaard claimed to be the penetration of eternity into time? Often when we evoke eternity, it is to cast it out into a kind of post-time, as if time were only the nasty sin of eternity. But we sometimes have the feeling that it dozes somewhere in the depths of the present. Aren't the events that mark us "for life" linked with the furtive rather than with the never-ending, and with radiance rather than consistency? Any "eternal instant" mysteriously intertwines the fleeting and the definitive, and steps back from the banality of physical instants, which are all the same.

The present imitates for us notions that are apparently opposed to it and are scarcely found in its physical representation. In the thirteenth century Saint Thomas Aquinas drew on a cyclical conception of time to explain the link, which at first sight was antinature, between the present and eternity based on the idea that: "Eternity is always present at any time or moment of time. We can see one example of this in the circle: a given point of the circumference, though indivisible, does not coexist with all other points, because the order of succession constitutes the circumference; but the center, which is not in the circumference, finds itself in an immediate relationship with any given point on the circumference. Eternity is like the center of the circle. Though simple and indivisible, it encompasses the whole flow of time and each part of it is also present."[4] Eternity would then be the pivot around which time turns, and each moment would be, against all expectations, filled with the infinite. Even if this explanation loses relevance in the frame of linear time, it suggests, in a metaphorical way, that the present instant is sometimes not foreign to the timeless.

But not all present instants are magical. Their existential density—what they represent to us—seems to go from zero to infinity. There is *brave time,* which rushes vigorously forward without regrets toward its own succession; *submitted time,* which lingers and laments; *poor time,* which reveals nothing except its mediocrity; the *petrified time* of melancholia, in which life tries to reverse the way it passes by (to go "uphill backwards" as René Char said); and the *compact time* of impatience, which substitutes for the present what it announces or promises. But there is also the *time emerging from passion,* which drives drunk, which leaps into existence with a sort of infinite momentum.

In short, if there is only one time, it is for us never the same: only the shimmer that our mind projects on it gives this vivid chameleon its deceptive colors.

23

the unconscious; or, time without flow

Pale sun of forgetfulness, moon of the memory
What do you drain at the bottom
of your deaf lands?
SUPERVIELLE

We know since Freud that consciousness is not the master of its own house. There is also the unconscious, for which time, ignorant of causality, seems to be reduced to scattered bones.

Concerning the unconscious, the father of psychoanalysis indeed elucidated a fundamental hypothesis about time and memory. According to him, things are simple: the unconscious ignores time. More precisely, the unconscious does not suffer from the effects of time, because it does not decline, does not weaken. Nothing ever cuts into its demanding power. It is something like the past—impossible to change. How does Freud draw this conclusion? From the following observation: what we discover in analysis—the symptoms, the formation of the unconscious—is not marked with any temporal clue, is not dated. He writes: "In the unconscious, nothing finishes, nothing passes, nothing is forgotten."[1] In a dream we can find thousands of details from a past event that we would have a hard time remembering when awake. Another argument: the same dream analyzed over the years will always reveal the same associations, as if no process (other than the "cure") would seem able to wear out marks from the past in the unconscious, even the remotest marks, to the point that we can claim that what has the oldest links in the past is what is the most determinant in the *psyche*.[2] The unconscious thus seems able to establish relationships with physical time, which are not the same as those of ordinary awareness. It dresses it up otherwise, even sometimes forcing it to change outfits.

In dreams the flow of time does not always wander from before to after. Freud says that sometimes "the dream shows us the rabbit chasing the hunter." Thus instead of thinking of time as a homogeneous and causal flow in which each instant has the same factual value, the unconscious stages a dismembered temporality: its diverse parts, instead of succeeding one another in a linear way, are in tension with one another; the overall structure is modified by various intensifications; different periods mix and juxtapose themselves like layers of lava infringing on one another, making the results of successive eruptions coexist. The unconscious makes do with keeping—outside of any linear temporal relationship—traces of the deposits of the past, which are retained only according to their intensity and can always contaminate one another, to the point of sometimes blurring the order of their succession: it is sometimes only after the fact that a past event actually becomes an event. A time-lag effect intervenes between the date of a past event and its assimilation by the subject.

Freud speaks about the *Nachträglichkeit,* which we can translate as the "capacity of happening later." Thus a certain situation, whose sheer complexity proves that an eighteen-month-old child could not perceive it, finds verbal expression at four years old but is not truly grasped until twenty years later. Thus a dream, lived as the pure present, can send the subject back to different periods in the past, from relatively recent events to early childhood, all expressed in dispersed order. It is not by arranging these events in a chronological way that the latent meaning of the dream can be revealed, because there is not *one* story but *several* stories that interweave, overlap, and sometimes contradict one another, each one living at its own rhythm according to its own temporality. Everything happens as if the unconscious recognized neither the flow of time nor the causality linked to it.

When we put all these elements together, the time of the unconscious actually appears like "shattered time," to quote André Green.[3] Even if there really is an invariance of the unconscious, it would be better to avoid calling it "timelessness." The fact that physical time, with its clearly defined flow, is not a form recognized by our psychic acts is one thing. But if the unconscious holds everything and never wears out, it is because it is itself carried by the flow of time, which keeps on *making it* identical to itself, anchoring it in time. The fact that it avoids the arrow of time does not at all imply that it is out of time. It merely lingers in a time without becoming. We can therefore wonder if Freud does not go a little too far when he claims that the "processes of the unconscious system are timeless, meaning that they are not temporally organized, are not being changed by the flow of time, have absolutely no relationship with time."[4] Because if it is really a matter of *process,* a kind of *procession,* how could they happen without "any relationship to time"?

Freud is more convincing when he develops the link between the unconscious and "the ability to forget." He first notices that certain repetitive behavior shows that circularities are at work in the unconscious. The subject repeats an act instead of remembering its first occurrence, reproduces it instead of remembering it, as if a short circuit had made him lose his memory of what his repetition is linked with and immobilized his actions in a sterile circularity. Freud says that such a subject repeats instead of remembers, but we could just as well say that he repeats in order *not* to remember. The more he repeats himself, the less he can remember, and the less he can remember, the less he knows why he repeats and then desperately tries to repeat so as not to take the risk of becoming aware of the meaning of what insists and reiterates in himself. The single memory that remains is of the impulse that pulses, rhythmically, endlessly.

This ability to forget is always valued less than remembering, as if it only expressed a failure of consciousness, a failure of memory. But such an ability is as important as memory, since it is that which, in the long run, clears the mind, soothes the affects, and protects us from the torments of the past. But who manages it? The unconscious, answers Freud without surprising anyone. According to him, in addition to regular memory, the one that records information, retains it, and makes it available later on, there is a specific memory that is peculiar to the unconscious. It is a "memory of forgetfulness," meaning that the events it records seem to be totally forgotten by the subject. This memory represses events, especially if they are decisive, by setting barriers that prevent it from remembering (and thus prevent it from really forgetting them).

This kind of memory, which doesn't suffer the effects of time passing by, is what gives the unconscious nucleus its unchanging and definitive character. We see a sort of paradox: what is most decisive, what is most strongly inscribed, what does not suffer time's normal wear and tear, unlike the memories that the subject easily recalls, is what appears to be "totally" forgotten.

A quantum physicist would say that this memory of forgetfulness is the "hidden variable" of our psychology. It secretly spreads invisible determinisms in our psychological makeup. It comes from the past but keeps affecting us in the present, speaking in us without our awareness of it.

24

the physicist, the romantic, and the jealous type; or, the dramas of impossession

I loved some beings; I lost them.
I became crazy when this struck me,
Because it is hell.

MAURICE BLANCHOT

Quickly, with his insect voice, Now says:
I am the Past and I sting you with my
Hideous thorn!

CHARLES BAUDELAIRE

Not everything goes away with time, we were saying earlier, since the physical laws that rule the world are supposed to rebel against history. The physicist easily admits that beyond the effectiveness it provides him, he has a very platonic way of using mathematics. The timeless truths that he draws from it seem to cause a certain nostalgia for fixity: doesn't mathematics double the reality that is there in front of him but changing and ungraspable, double it with an explanatory behind-the-scenes world filled with imperishable entities?

If the physicist uses only strategies that wear scents of eternity, it is because the reality he tries to grasp always eludes him: he manages to understand the immediate reality only if he considers it first as the expression of another reality, which would be perfect and unchanging. A kind of "drama of impossession" unfolds, to use Clément Rosset's words: the certified evanescence of reality is being "sublimated" by the invocation of an explanatory timelessness that is judged to be more fundamental.[1]

This approach does not make the physicist especially original. The same fascination for the unchanging is found in various philosophical or human enterprises, whether it is to avert the passing of time, to capture elusive data, or to understand an ever-changing reality. This is not a chance similarity, but a necessity: in order to understand the world, we first have to discern some "prehensible"—that is, fixed—entities. In sum, ideas and concepts seem to have value to us only if we can deduce them from an invariable source that allows us to "grasp" them intellectually.

From this point of view, the physicist reminds us of the romantic or the jealous man, in that all three of them express a certain distress of impossession. For the romantic, this comes not from what time withdraws immediately after giving, but what it never gives itself, just as the impossession felt by the physicist comes from the fact that the real keys for explaining the world are not directly perceptible in the world itself. But the romantic drama is renewed, by construction, at each moment throughout life, because no moment is graspable but all of them consume us. The instant cannot be captured *even though it is right there,* within our grasp. The present progressively effaces itself from under our steps as it passes. Does this explain the symbolic importance we grant to commemorations and anniversaries of all kinds? Does it explain our attempts to install artificial immobilities in the heart of a time that is fleeting? Rituals are comforting. Hence our concern about traces, our legacy, our creations, all these paths by which we human beings try, in vain, to persuade ourselves that we have some power over time. Hence, finally and foremost, the celebration of immortal love as a parade into destructive time: "What I loved one day, and kept or not, I will always love," claims André Breton.[2] Even if time gives itself to us, we heroically claim that it will never be able to deny the truth of the events and feelings that it made happen.[3]

Another drama of impossession, another nostalgia of fixity, of what has stopped, is amorous jealousy. It is certainly not by chance that Marcel Proust has analyzed the relationship with time on the one hand and the relationship with the beloved on the other with the same acuity. Aren't they similarly structured? Jealousy in love arises not only because the beloved is not ours, but also because he or she cannot be apprehended in him or herself: since the heart cannot own what it can only love, we only know the other in ourselves, with our imagination (to pursue Proust[4]) as our only scalpel.

When it is a question of drama, all accumulations are possible. We thus have to be able to find excellent physicists who are also very romantic and furiously jealous, and who, in addition, never win at cards.

25

has physics forgotten death?

I'm not afraid of dying.
I just don't want to be there when it happens.

WOODY ALLEN

You have little time.
Live as if you were on a mountain,
Here or there, it does not matter.

MARCUS AURELIUS

In any exercise of the intelligence, we make an effort to contest or defy the passage of time. As Spinoza explained with acuity: "It is in the nature of reason to perceive things through a certain eternity."[1] The intelligible and the eternal seem to have always been associated. But this kind of association has been evoked most often where art is concerned; there was a time when any kind of perfection on an aesthetic level had to be bound with eternity in some essential relationship.

In this matter, science has imitated the arts: it wanted to associate perfection and permanence. Where does the confusion come from? Here again Galileo puts us on track: according to him, our taste for stability comes simply from our dreadful fear of death. He had this idea after he renounced the Aristotelian distinction between the local world, supposedly imperfect and corruptible, and the distant world, supposedly perfect and incorruptible, made of an unchanging "quintessence." Having aimed his telescope toward the sky, he discovered the uneven surface of the moon, "uneven and covered, like the surface of the Earth, with high hills, deep valleys and crevices."[2] Matter was thus the same everywhere—"earthy"—here, on the moon, or anywhere else. Since it was obeying the same laws everywhere, since it was as degradable in the sky as on earth, only one kind of universe was to be considered, composed of only one kind of matter, which was perishable everywhere. Galileo deduced that we had been wrong to associate permanence and perfection: what is perishable does not have to be imperfect.

For Galileo, the cause of this metaphysical amalgam was clearly designated: "I think that people who praise incorruptibility, permanence, end up saying these things because of their great desire to survive and because of their fear of death. . . . And it is beyond doubt that the Earth is much more perfect, as it is, alterable, changing, as if it were a mass of stone, or even just a very strong and impassive diamond."[3] Would death be another costume of time, the most discreet, but also the most deceptive of all? Its ultimate underwear, in a way?

The question is real since time seems to be simultaneously what makes things last and what makes it that nothing lasts forever: we last, we last, and one day we stop lasting. Every death refers to the end phenomenon (something stops), but also to the end of phenomena (we don't know what happens next), thus combining nothingness and the unknown. Hence, death is simply an infinite mystery. We understand corruption, transformation, dissolution, and we grasp the idea that something can live on while forms fade away, but death settles on all of this and remains resistant to thought, science, or speech.[4] Yet physics seems unable to say anything about death. Too concerned with unchanging laws and perennial relationships, has it forgotten death? Or has it brushed it aside?

There is a radical difference between physics and biology that Georges Canguilhem underscored: "The sickness and death of these human beings who produced physics, sometimes risking their lives, are not problems of physics. The sickness and death of physicists and biologists are problems of biology."[5] As a matter of fact, physics has limited its ambitions and marked out its domain. It studies only inert matter and assumes that all material objects that it identifies—atoms, for example—are not alive, even when they belong to a living being: wherever they are, they are lifeless entities, only their numerous and organized agglomeration has been able to produce life. Life would thus be just an emerging property of inert matter. This hypothesis is not shocking; after all, a group of atoms usually has properties that individual atoms do not have (the atoms that constitute a red painting are not red).[6]

Thanks to these kinds of arguments, we have the right to consider that living matter and inert matter, despite their apparent differences, are ruled by the same physical laws. We can even add that because of the very peculiar organization of living matter, the implementation of physical laws produces very specific results in it. Only the circumstances of their implementation change. But without returning to vitalism, it is clear that the claim to describe life only by dissecting the lifeless objects that contribute to it might be a little simplistic. The gene, the molecule, and the atom are indeed three entities that take part in life, but this knowledge, as elaborate as it can be, is not the knowledge of life. Life seems to have a kind of extraterritoriality—if not by principle, at least de facto—compared to physics. The mechanistic approach, which sets aside and favors the material substratum alone, misses the living. But a dilemma emerges: how can we study the living in itself, independently from matter?

In a "reciprocal way" it is rather easy to find reasons why physics is not at ease with the living; three of them are cited here. The first is linked to mechanics (the first historical success of physics), founded on the inertia principle, which protects the movement of matter from the executive power of life. Inertia is inactivity, permanence, indifference. It is also the neutrality of time: movement keeps on being what it is, unless a force appears and changes it. Life is the polar opposite of an indifferent relationship with time and surroundings, to the point that a philosopher as rigorous as Kant identified the inertia of matter with the absence of life.[7] We will retort: "Yes, but radioactivity reintroduces temporality in lifeless matter!" Of course, radioactive atoms end up "dying" by disintegrating into other particles, but saying that their death is the result of an aging process is a real leap—a leap we should not take. Indeed, the probability they have to disappear in a given time gap is strictly independent of their age: a three-thousand-year-old carbon-14 atom and a five-minute-old carbon-14 atom have the exact same probability of disintegrating in the following hour. Their disappearance is thus not the result of any structural alteration: they die having aged, but without growing old.[8] On the contrary, it seems that in living systems, exchanges with the exterior become less efficient over time; cell renewal slows down, as if there were a certain wear and tear on the mechanisms in action. It is at least in these terms that the first theories of biological cellular aging were expressed, which have the merit of underlining one essential point: the augmentation of the death rate with age. Contrary to radioactive atoms, our probability to die changes with time. This is even the exact meaning of the word "aging."[9]

The second reason comes from the fact that, as we saw, according to contemporary physics, all phenomena occurring at a microscopic level are reversible, which means they are indifferent to the flow of time; everything that is done can be undone. At this scale, the flow of time does not cause anything ineluctable—no scrapes, no wrinkles, no death—so nothing seems to be aging.

The third reason, certainly the most fundamental, is linked to the fact that physics always tries to look for invariable relationships between phenomena, relationships beyond the reach of change. Since its natural tendency consists of the desire to express the future starting with elements that escape the future, in telling stories with story-less rules, we realize that physics has a hard time explaining a phenomenon as violently discontinuous as death.

For these reasons, extending the methods of physics to the study of aging—and of the living in general—can only trigger fierce resistance. This resistance, which expresses an affective reticence toward any mechanization of the living, especially expresses skepticism toward a paradoxical hope, that of explaining a phenomenon using laws that are based on a hypothesis that denies them.

There are some leads nonetheless. We have already evoked the fact that certain spontaneous physical changes happen more as destruction and disorder; any group of objects, whether it's atoms or galaxies, actually tries to occupy the maximum space at its disposal, taking into consideration the interactions that exist between these objects. The structures initially organized within it eventually disappear. This is at least what the second principle of thermodynamics forecasts when it says that the entropy of a *closed* system can only grow.

But this principle cannot apply to living cells, because they are, like all living organisms, open and not closed systems: they exchange matter and information with their environment, incorporate exterior elements, renew their substance, which is thus younger than the organism itself, react to aggression, and spontaneously heal from certain diseases. They are in a permanent struggle against the destiny of universal disorganization (when it is applied without caution) that the second principle of thermodynamics promises them.

Physics was thus right to limit its ambitions. First and foremost it was right because it knew how to convert its modesty into a power of intelligibility, on a scale going from quarks to galaxy clusters. It was also right because, as we just pointed out, it loses its arrogance when we want it to intervene outside the field to which it had first been confined. But we should not reach premature conclusions: the difficulties we have evoked could be solved one day, thanks to the growing number of bridges built between physics and biology. It is better to consider that the question of knowing whether the specificities of living beings are ultimately reducible to "enlarged" physical laws remains open. But this also means accepting the fact that as long as this question is not solved, death will remain physics' great "unthought." We will have to make do without physics to explain the link that death weaves with time.

Apparently, death happens as an effect of time. It dresses time with finitude. It is then an easy leap to think that time is, if not death itself, at least its vector, that death determines its structure and that we have to think of time starting from death and not the reverse—an easy leap that many philosophers have dared to take. Doesn't the perspective of death inhabit life itself? Doesn't it set in motion, *for us*, a finished temporality? Certainly, and Montaigne had seen this clearly: "You are in death while you are alive." But Heidegger radicalized this idea to the extreme. According to him, death is the source of all of our ordinary representations of time, for the simple reason that death precludes locating time in a larger order.[10] Thus time would simply be the other name of death, a less stressful name, more neutral, the ultimate trick through which we manage to reduce the affective power of the word "death." Time would be nothing more than a mask of death, more vivid than death, destined to make it verbally presentable and intellectually admissible.

This conception has no lack of arguments. The idea of death has an unquestionable impact on our human time, more exactly on our human perception of time: death gives time a peculiar pine tree scent,[11] a diffuse perfume that impregnates all our thoughts of time, as if we could not think of it without considering the unstoppable destruction it promises us. Even though we know it is not the end of time, but only the end of the time of one being in the uninterrupted flux of time, death stands in the way. Of course, we have several more or less efficient strategies that enable us not to feel its shadow too much. Intoxication, as suggested by Baudelaire, is one of them.[12] Other strategies are to have children, who will then also have children; acquire notoriety or glory; anaesthetize ourselves with multiple activities; climb incredible summits; win at the Olympics; invest in buildings; or if you are a rich man, get married to a very young woman when you are old. All of these help us to temporarily forget, in the illusion of lasting almost infinitely or to persist by proxy, to face our mortal fate.[13] But death always wins, hands down. No deception ever beats it. Its records are full, always up to date. It never forgets anybody, whether nabob or slave, arrogant or humble, as if a nice communism from beyond the grave had to contrast the equality of corpses with the inequality of living beings.

How should one behave when faced with this unavoidable temporal wall? Limitless fear, indignation, making a big fuss at the idea of not seeing the sunset; freezing, shrinking up, and thus dying before being dead; living as if we should never die, writing blank checks as if eternity were at our disposal, "knick-knacking" death by imagining it withdrawing to a very remote corner of the sky or entrenched in a cave on the other side of the world?

But I can just as well find it sweet to tell myself that one day I will no longer exist, and consider this new morning as a new grave given to me. Doesn't every lived instant, as soon as it detaches itself from the dark background of death, immediately get some sparkle? By keeping life in finitude, death abandons us to our compulsory glory. It makes us precious, pathetic, touching: every act can be our last; every face threatens to disappear in the following second. But the future is not reduced to the imminence of death. It is only a moment of the future, not the shroud of the present. Thus, rather than thinking of time according to death, we should think of death according to time and for what it is: an event to come *in* time.

We would be better off keeping to a diet of the passing moment, trusting the flavor of the instant, the *kairos. I am going to die. So be it!* It is thus the perfect moment to enter existence, to colonize the ephemeral. Night will be here soon enough. These hours, these minutes, these seconds are events. Derisory events? Maybe. But this "derisory" was, is, and will be exactly me.

We must learn to love the irreversible.

notes

introduction

1 When they interact with matter, the protons emit X-rays, whose energy spectrum provides information about the nature and concentration of different chemical elements present.

2 To upset Aristotle's theory, Galileo certainly did not need to drop things from the top of the Tower of Pisa. All he had to do was notice, thanks to some reflection, the theory's intrinsic contradictions. According to Aristotle, if you took two balls—a big one and a smaller, lighter one—and dropped them, the big ball would fall faster than the small ball. If we were to tie them together, though, the two balls together would be heavier than the big ball alone and would fall *even faster* than the big ball would. But we could also deduce from Aristotle's law that the small ball would break the fall of the big ball, so that together they would fall *slower* than the big ball. These two arguments drawn from the same law thus reach two contradictory conclusions (which is a royal achievement for the father of the law of the excluded middle). The only way to eliminate this paradox is to affirm that all balls fall in the same way, no matter their mass.

3 This expression, attributed to Minette, the mysterious Vaudeville dancer, is picked up by Lousteau, a character in Balzac's *Lost Illusions (a novel in three parts, published in 1837, 1839, and 1843).*

chapter 1

1 Heidegger, Martin. *Being and Time*. Transl. Joan Stambaugh. New York: State University of New York, 1996.

2 The English loses the pun of the original French, which reads, *Le temps loge hors de l'horloge.*

3 When a certain amount of time had passed, the hammer fell on a bronze bell with great ruckus. But these timepieces had the inconvenient aspect of all water clocks: their flow rate varied with the temperature and atmospheric pressure, not to mention freezing conditions, leading the monks to unexpectedly sleep in. As for the first hourglasses, which appeared in the fourteenth century, their precision left much to be desired. Even the finest sand ultimately eroded the glass container and thus accelerated its own demise. For many long years, hourglasses were limited to regulating guard duty for fortified castles, and later they became the worldwide standard for cooking perfect hard-boiled eggs.

4 Between the thirteenth and fifteenth centuries, *none* stood at noon, *vespers* at the beginning of the afternoon.

5 For more about this, see Norbert Elias, *Du temps,* trans. Michel Hulin (Paris: Fayard Press, 1997).

chapter 2

1 "Time is this way. Who can define it? And why attempt to do so, since all men know what is being spoken of when we talk about time without any added definition." Blaise Pascal, *Pensées,* "De l'esprit géometrique." Ed. Garnier Frères (Paris, 1960). 2. Michel de Montaigne, *Essays,* book III, chapter 13, ed. Pierre Villey (Paris: PUF, 1978), 1069.

3 Saint Augustine, *Confessions,* book 11. Trad. L. de Mandadon (Paris: Seuil, 1982).

4 Martin Heidegger, *Acheminement vers la parole* (Paris: Gallimard, 1976), 200.

5 A Chinese proverb says more or less the same thing, "For men, time is passing; for time, men are passing."

6 Wittgenstein gave numerous variations of considering things this way. For example, proposition 268 in *Of Certitude* says, "I know that this is a hand. And what's a hand? Well, *this,* for example."

chapter 3

1 Marc Aurèle, *Pensées,* book IV, 43, in *Les Stoïciens,* trans. E. Bréhier (Paris: Gallimard, coll. "Bibliothèque de la Pléiade"), 1166.

chapter 4

1 See Jean-Pierre Vernant, *L'Univers, les Dieux, les Hommes: Récit Grec des Origines* (Paris: Seuil, 1999).

2 Before becoming an ignominious tyrant who ate his children as soon as his wife, Rhea, gave birth, Kronos was a liberator. Meditate on this one with the Kantian background music of "Every liberation is not a deliverance."

3 Satapatha Brahmana, 11, 1, 6, quoted in "*Les Mythes de la Création,*" in Frédéric Lenoir and Ysé Tardan-Masquelier, eds., *Encyclopédies des Religions du Monde,* book 2 (Paris: Bayard-Centurions, 1997), 1523–24.

4 But as Ernst Cassirer explained in his *Essai sur l'Homme* (Paris:

Editions de Minuit, 1976), this does not mean that myth is a "failed science." Myth is a symbolic form that offers an explanation of the world, and more precisely an explanation of the world's *meaning*. By describing the genesis of the world, it tries to provide a truth. Science, however, is not concerned a priori with the question of meaning.

chapter 5

1 Erwin Schrödinger, *Carnets de 1919. A propos de Philosophie Kantienne,* quoted by J. Mehra and H. Rechenberg in *The Historical Development of Quantum Theory* (New York: Springer Verlag, 1987), 40.
2 Jean Bernard, *Le Jour où le Temps s'est Arrêté* (Paris: Odile Jacob, 1997), 11.

chapter 6

1 Twenty-five centuries seems to be so far away that we do not see which thread could tie us to these pre-Socratic thinkers. But there is a very visible one, which Max Dorra noted: in Sequoia National Park, nearly equidistant from Los Angeles and San Francisco, stands the General Sherman tree, a 275-foot sequoia that is estimated to be over 2,600 years old. It was just a little sapling in the days of Parmenides and Heraclitus (see Max Dorra, *Heidegger, Primo Levi et le sequoïa* [Paris: Gallimard, coll. "NRF," 2001]).
2 Echoing fragment 95: "Everything goes away and nothing remains." (Heraclitus, *Fragments* [Paris: GF-Flammarion, 2002].)

3 In the same way, the invariance of physical laws under space translation—meaning they are the same anywhere—has as a consequence the impulse preservation. This preservation law particularly forbids any spontaneous change of movement, according to the inertia principle. This is the same as saying that space is homogeneous, that its properties cannot change from one point to another.

4 Particle physicists are now using "colliders," particle accelerators in which two beams move in opposing directions at nearly the speed of light and can collide. In these clashes, all energy from incident particles is convertible into mass, according to the equivalence between energy and mass ($E = mc^2$). This way it can fully transform into other particles.

5 P. A. M. Dirac, "A New Basis for Cosmology," *Proceedings of the Royal Society,* Series A, Mathematical and Physical Sciences, vol. 165, no. 921 (April 1938): 199–208.

6 The fine structure constant is not a fundamental physics constant, but an adimensional combination of the Planck constant, the light speed, and the electron charge. Some recent theoretical models, including the superstring theory, do not rule out the fact that the fine structure constant might have changed at a certain level (as a matter of fact, the superstring theory predicts a priori variations in this kind of constant). The best way to check if this variation is real or not, and to detect in general a possible variation in other constants, involves testing the "equivalence principle" of general relativity with very high precision. The "weak" side (Galilean) of this principle simply states that an object's movement through a gravitational field is independent from the object's internal structure or composition; in a void, one pound of lead must fall like one pound of zirconium. The

"strong" formulation, proposed by Einstein as a founding principle of general relativity, stipulates that the physical laws remain locally identical in the absence or the presence of gravitation. Thus, an observer locked in a falling elevator without windows—or in a space capsule with shutdown engines—cannot experience anything that reveals to him the external presence of the masses toward which he is falling. Locally, the gravitational effects are indiscernible from the acceleration effects. Several experiments on satellites are preparing to test this principle with precision.

7 J. M. A. Paroutaud, *La Ville Incertaine* (Paris: Le Dilettante, 1997).

8 *Le Figaro,* December 29 and 30, 2001, p. 3.

chapter 7

1 Gustave Flaubert, *Correspondance, p.* 87 (June 7, 1844) Ed. By Jean Bruneau (Paris: Gallimard, 1998).

2 Christian Bobin, *Ressusciter* (Paris: Gallimard, 2001), 71.

3 According to some historians, however, the first philosopher who really asserted the existence of physical time was Albert Le Grand (1200–1280), Thomas Aquinas's teacher. Taking the opposite view of Saint Augustine's thesis, Le Grand asserts that time really exists in nature and that the soul only detects it: "*Ergo esse temporis non dependet ab anima, sed temporis perceptio,*" which can be translated as, "What depends on the soul is not the existence of time, but the perception of time."

4 The ancestor of the sundial, the gnomon is the simplest instrument to measure the passing of time. A vertical post in the sunlight projects a shadow whose length changes according to the time of day (it is shorter at noon, when the

sun reaches its highest point of the day). But since this length changes during the year as well, the meaning of the marks on the gnomon's horizontal level is constantly changing.

5 Galileo did not exactly characterize physical time, nor did he try to define it explicitly. The properties of physical time are simply and implicitly determined by the nature of some of Galileo's laws. Thus when he studied uniform movement, Galileo established a geometrical relationship between the notions of space and time, restricting their common link: "By regular or uniform movement, I mean all spaces covered by a moving object in equal times are equal to each other." In his *Principia,* Newton really characterized physical time: "Absolute time, true and mathematical, with no relation to the outside, flows uniformly and is called *duration.*"

6 François Jullien, *Du "temps." Eléments d'une philosophie de vivre* (Paris: Grasset, 2001).

7 See *L'Espace et le temps aujourd'hui,* Emile Noël et Gilles Minot (dir.) (Paris: Seuil, coll. "Sciences," 1983), 273–87.

8 Henri Bergson, *L'évolution créatrice,* in *Oeuvres* (Paris: PUF, 1970), 503.

9 Henri Bergson, *Essai sur les données immédiates de la conscience,* in *Oeuvres* (Paris: PUF, 1970), 85.

10 Quoted by Maurice Merleau-Ponty in *Signes* (Paris: Gallimard, 1960), 248.

chapter 8

1 The variable *t,* which represents time, is what we call in mathematics a *real* number.

2 Bergson, Henri, *Essai sur les données immédiates,* 1889, 76–77.

chapter 9

1 There is no better study of the circle's symbolism, including literature, than Georges Poulet's (*Les Métamorphoses du cercle* (Paris: Flammarion, coll. "Champs," 1979).

2 For more about the subject of time in myths, see Mircea Eliade, *Le mythe de l'éternel retour* (Paris: Gallimard, coll. "Folio Essais," 1989).

3 As for mixing time and movement, the apparent circularity of time prosaically exposes itself in the movement of a watch's hands. This round representation has indisputable tranquilizing powers. Before committing suicide, Jean-Louis Bory explained that he couldn't bear the materialization of the flow of time with the appearance of the quartz watch; the hands traveling around a closed circle seemed less harrowing to him than the scrolling by of the seconds, minutes, and hours whose end could only be death, including his own.

4 The discovery of irrational numbers quickly made this idea impossible to defend, since the revolution periods of various stars, which shared the "big year" as the smallest common denominator, could no longer be expressed by whole numbers.

5 See Odon Vallet, *Petit lexique des idées fausses sur les religions* (Paris: Albin Michel, 2002), 244–45.

6 A thousand "divine" years corresponds to millions of human years.

7 Using statistical physics, we can demonstrate that each classic system evolving according to determinist laws ends up in a state as close as possible to its initial state, after a long or short, but never infinite, duration. This is Poincaré's "recurrence theorem," as demonstrated in 1889.

8 Friedrich Nietzsche, *Oeuvres,* vol. X, trad. Jean Launay (Paris: Gallimard, 1978), 20–21.

9 Ecclesiastes I:9 (italics in original).

10 Hence Schopenhauer's interest in historical studies that make you become aware of the same thing, hidden under the illusions of change, and that simultaneously denounce the manifestations of a necessarily illusory evolution, which is only passing old dishes by.

11 In Greek mythology Sisyphus is condemned to repeatedly push a rock up a hill. Upon reaching the top, the rock immediately rolls back down and Sisyphus has to begin his task again. He does so eternally, wearing out his legs and nerves. If we have to "picture a happy Sisyphus," as Albert Camus says, we cannot forget that his destiny resulted from a punishment dealt by the gods for having dared defying death.

Also in Greek mythology, Ixion, a king judged guilty of murder and betrayal, is purged by Zeus, who pities his distress. He is even invited to Mount Olympus, where he drinks the nectar and ambrosia, thus becoming immortal. Completely unscrupulous, he tries to seduce Hera, Zeus's wife. Furious, Zeus shapes a cloud in the image of his wife, and Ixion mates with it. This illusory union gives birth to the centaurs, horses with a human torso and head. Convicted for his ingratitude, Ixion is tied to a burning wheel that turns eternally in Tartar, a place that is synonymous with hell. This is a horrible torture, since eternity, as Franz Kafka (but also Pierre Dac and Woody Allen) says, "is long, especially close to the end."

12 Nietzsche, *Oeuvres,* vol. XII, 213.

13 For more on this subject, see Gilles Deleuze, *Nietzsche et la philosophie* (Paris: PUF, 1962), Clément Rosset, *Le choix des*

mots (Paris: Editions de Minuit, 1995), or Jean Brun, *Heraclite ou le philosophe de l'éternel retour* (Paris: Seghers, 1965).

14 Quoted by Jean-Jacques Rousseau, "Time" from Diderot and d'Alembert's *Encyclopédie,* t. XVI, 1765.

chapter 10

1 Leibniz, Gottfried Wilhelm, *Principes de la nature et de la grace fondées en raison* (Paris: PUF, coll. "Epiméthée," 1986), 112 (original text in French).

2 When Marechal de La Palice died in the battle of Pavie in 1525, his soldiers said, "Fifteen minutes before his death, he was still alive." Thus a statement of the obvious, a *lapalissade* in French, comes from the name La Palice.

3 This makes the causality principle impossible to explain from a strictly philosophical point of view, since causality is the principle from which everything comes; causality itself proceeds from nowhere (causality is incapable of explaining itself, since it doesn't have a cause). This statement pushed Schopenhauer to run the opposite experiment of Newton, who wondered why the apple fell and discovered the reason why. Schopenhauer was surprised that causality, as discovered by Newton, was enough to make an apple fall and wondered how causality had become an obvious fact for so many philosophers. The problem is simple: do the questions stop being asked when it is said that "an apple falls toward the earth because of the law of gravity," or should we additionally answer by searching for the cause of gravitational force itself? The superstring theory, which will be discussed later, might answer this last question.

4 But the cause concept remained present in thinking. As

Bertrand Russell noticed, the cause concept, like the English monarchy, was left alive only because we assume that it's harmless. (Bertrand Russell, *The Concept of Cause,* in *Mysticism and Logic* [London: Allen and Unwin, 1986], 173).

5 The "wave function" of a particle is a mathematical function from which we can calculate the probability that the particle appears here or there.

6 See Thomas Kuhn, "Les notions de causalité dans le développement de la physique," in *Les Théories de la causalité* (Paris: PUF, 1971).

7 In a very well-argued book, Max Kistler defines causality, in the modern sense of the word, in this way: "Two events *c* and *e* are linked as cause and effect if and only if there exists at least a physical magnitude P, subjected to a law of conservation, exemplified in *c* and *e,* and whose definite quantity is transferred between *c* and *e.*" (Max Kistler, *Causalité et lois de la nature* [Paris: Vrin, 2000], 282.)

8 Thiery Jonquet, *La Vigie et autres nouvelles* (Paris: L'Atalante, 1998). This short story has been adapted as a comic book (Jean-Christophe Chauzy, Thierry Jonquet, *La Vigie* [Bruxelles: Casterman, 2001]).

9 Jonquet, *La Vigie et autres nouvelles,* 150.

chapter 11

1 A tachyon is defined as a particle that travels faster than light in space. If such an object existed, it would make time travel possible. For this reason, the theory of relativity excludes this possibility by virtue of the principle of causality.

2 Criticizing this H. G. Wells novel, Alain insightfully writes: "The observer who drove the machine comes back to the time

he left, finds his friends and the universe the way they were at the beginning. This means that there have to be states of the universe existing at different times, which doesn't work at all. I'm not refuting this novel, which is beautiful, but I want to make this time condition a little clearer, showing that all things travel along it together and at the same pace." (Alain, *Propos* (1923), in *Vigiles de l'esprit* [Paris: Gallimard, 1942], 245–46.)

chapter 12

1 Jean Eisenstaedt gives an excellent presentation of general relativity in his book *Einstein et la relativité genérale* (Paris: CNRS Editions, 2002).
2 Gödel presented his model to Einstein, who was not convinced, explaining that he was unable to believe that we could "send telegrams to one's past."
3 This matter would then behave, from gravity's point of view, in opposition to ordinary matter, in that it would be able to defocalize light beams with gravity.
4 Kip Thorne, *Trous noirs et distorsions du temps* (Paris: Flammarion, coll. "Champs-Flammarion," 1997), 518–59.
5 For more about this subject, see Gabriel Chardin, *Peut-on voyager dans le temps* (Paris: Le Pommier, coll. "Les petites pommes," 2002).
6 Arthur Rimbaud, "Mauvais sang," *Une saison en enfer* in *Oeuvres complètes* (Paris: Gallimard, "Bibliothèque de la Pléiade," 1972), 96.
7 The fixity of time has always seemed so etched in stone that even the most daring theologians have limited divine omnipotence by refusing to grant God the exorbitant power

to erase or change the past as He likes, to rewrite history, to thoroughly review the world, because this power seems inconceivable. For Descartes, even God in his "most powerful" version cannot avoid the fact that what has been, has been. Thus any event in the past remains eternally true.

8 More precisely, and for those who know the lingo of theoretical physics, causality is expressed thanks to commutation relations between field operators. An operator of creation, $\Phi(x)$, of one particle at the space-time point *s* and the annihilation operator of this very same particle, $\Phi(y)$, at the space-time point *y* have to commute for a separation of the time genre: these rules prevent a particle from spreading on a line of the space genre (meaning that the particle would spread faster than light) and for the propagation on a gender timeline, that the particle creation happened before its annihilation. These rules can be satisfied only if the decomposition in flat waves of the field operators has some negative wave modes. What should be done with these modes, which in quantum physics correspond to negative energies, that is, particles that go back in time? We reinterpret them as *antiparticles* that follow the normal flow of time. Particles and antiparticles have to have the same matter and opposite electric loads. The antiparticle concept, and antimatter in general, is the price to pay so that formalism of particle physics (what we call "quantum field theory") can be compatible with relativity and causality. To learn more about this concept, see Gilles Cohen-Tanoudji, "Le temps des processus élémentaires," in *Le temps et sa fleche,* Etienne Klein and Michel Spiro (dir.), (Paris: Flammarion, coll. "Champs," 1996), 93–130.

9 The antiparticles are, as a matter of fact, mathematically equivalent to entities for which time passes backward, from the future to the past. This result, demonstrated by the

Swiss physician E. C. G. Stückelberg, was taken very seriously by Richard Feynman and formalized in his eponymous "diagrams."

10 At first, Dirac preferred saying that these new negative energy particles corresponded to protons rather than to positive electric electrons. As a matter of fact, in 1930 the proton was, along with the electron, the single charged elementary particle known, and Dirac did not want to introduce a new entity that nobody had observed.

11 Actually, when Carl Anderson discovered the positron, the antiparticle associated to the electron, he didn't know that its existence had been predicted by Paul Dirac. At first he granted it a mass equal to one-twentieth of the electron's mass, whereas Dirac's equation imposes that the antiparticle has the exact same mass as the associated particle and an opposite electrical charge.

chapter 13

1 They can do it, for example, by exchanging light signals. According to Newtonian physics, light travels at an infinite speed, so that this operation happens instantaneously.

2 This is true at least as long as the plane is flying at a constant speed and along a straight line.

3 More precisely, the vibration comes from the electromagnetic waves that constitute it.

4 Galilean referentials are referentials in rectilinear translation and all uniform. Relativity, whether from Galileo or Einstein, makes them all equivalent, because the laws of physics are expressed in the same way.

5 The muons are a kind of heavy electron naturally produced in the upper atmosphere by cosmic radiation.

chapter 14

1 André Comte-Sponville, *Dictionnaire philosophique* (Paris: PUF, 2001), 77.

2 Einstein himself wrote in his private correspondence (a letter written March 21, 1955, to the family of his friend Michel Besso, after his death) that, "for us convinced physicists, the distinction between past, present and future is only an illusion, even if it is persistent." Even if his point of view on the question has not always been that radical (maybe he just wanted to soothe the family of the deceased with this sentence), Einstein was still hoping to bring physics back to a pure geometry, meaning formalism without history.

3 Thibault Damour and Jean-Claude Carrière, *Entretiens sur la multitude du monde* (Paris: Odile Jacob, 2002), 52.

chapter 15

1 Some authors think we can define a "timeless causality" with two by-products: the flow of time and the arrow of time. If they are right, this would mean that physics could grasp the engine of time. See Marc Lachièze-Rey, *Au-delà de l'espace et du temps, la nouvelle physique* (Paris: Le Pommier, 2003).

2 The reversibility of microscopic equations has consequences on the macroscopic level through the so-called Onsager relations, which play on the possible exchange between causes and effects. For example, the coefficient that rules the heat wave induced by a gradient with electrochemical potential (Peltier effect) equals the coefficient that rules the electric wave created by a thermal gradient (Seebeck effect). It is possible to demonstrate that such symmetries reflect the absence of the arrow of time on the microscopic level.

3 Ilya Prigogine, *La fin des certitudes* (Paris: Odile Jacob, 1996), 126.
4 We can cite the remarkable experiments conducted by the Ecole Normale Supérieure's Physics Department under Serge Haroche's guidance (see Serge Haroche's article, "Entanglement and Decoherence Studies with Atoms and Photons in a Cavity," in the monograph *Physics of Entangled States,* Proceedings of the IX Seminaire Rhodanien 2001, ed. Robert Arvieu and Stefan Weigert [Paris: Frontier Group, 2002], 75–92).
5 A complete (but rather technical) list of all the arguments can be found in H. Dieter Zeh, *The Physical Basis of the Direction of Time* (New York: Springer-Verlag, 4th ed., 2001).

chapter 16

1 In practice, the variable T is changed into its opposite $-T$.
2 What is referred to as β radioactivity is the process through which, in an atomic nucleus, a neutron transforms itself into a proton by sending out an electron and an antineutrino.
3 The neutral kaons are short-term particles composed of a strange quark and an antiquark.
4 All mesons are formed by a quark and an antiquark. The "beautiful mesons" get their name not from any esthetic prerogative, but because they are composed of a quark or a "beautiful" antiquark, each owing their name only to purely contingent arguments.
5 Andreï Sakharov asked himself the following question: under what conditions is it possible to build a universe principally composed of matter starting from an initially symmetric universe, meaning that it first contained as many particles as antiparticles? He demonstrated that three conditions need

to be met. The first is the nonconservation of the "baryonic number," which is defined as the difference between the number of quarks and the number of antiquarks. No experimental evidence of this possible nonconservation has been found, despite in-depth research on the possible disintegration of the proton (the proton, being the lightest baryon, has to be stable if the baryonic number is conserved or, in the opposite case, must be able to break down into nonbaryonic particles). Sakharov's second condition is the violation of *C* or *CP* enabling us to distinguish matter from antimatter. The third condition is that the universe is thermically unbalanced, which "tips the scales" in favor of matter.

chapter 17

1 Gilles Châtelet, *Vivre et penser comme des porcs* (Paris: Gallimards, coll. "Folio," 1998).
2 Several tools have been used to study Type Ia supernovae, including the Hubble space telescope.
3 This team compared observable structures in cosmic background noise, the fossil radiance emitted 300,000 years after the big bang, with observable structures in the 250,000 galaxies gathered in clusters that have been listed in a systematic observation program. These results lead us to attribute the cosmological constant with a positive value.
4 This was not at all unreasonable, since the relative speeds of stars known at this time were all very slight and we were still unaware of the existence of galaxies other than our own.
5 In 1931 Einstein wrote an important article in which he explains that the discovery of the expansion of the universe by Hubble renders obsolete the cosmological constant that

he had initially introduced in his general relativity equations. This article was published in the *Sitzungsberichte der Preussischen Akademie der Wissenschaf*, and its exact reference is "Einstein. A. (1931). Sitzungsber. Preuss. Akad. Wiss. 235–37." However, it is obvious that numerous authors have quoted this article without having read it, so that, as in the game of "Telephone," the successive copies have progressively modified the references, introducing mistakes to the point of creating a totally unknown coauthor. We find, for example, in the material on the subject the following versions of the citation: "Einstein. A. (1931). Sitzsber. Preuss. Akad. Wiss. 235–37"; "Einstein. A. (1931). Sitsber. Preuss. Akad. Wiss. 235–37"; "Einstein. A. (1931). Sber. Preuss. Akad. Wiss. 235–37"; Einstein. A. (1931). Sb. Preuss. Akad. Wiss. 235–37"; "Einstein. A.S.-B. Preuss. Akad. Wiss. 1931. 235–37"; "Einstein. A. S. B. Preuss. Akad. Wiss. 1931. 235–37"; and finally, "Einstein, A., and Preuss, S. B., (1931), Akad. Wiss. 235–37." No doubt one day a science historian will get interested in the strange case of the young physicist "S. B. Preuss," who wrote only one article, but a major one, before disappearing from the stage. Heraclitus was eventually right: it is really impossible to repeat things identically!

6 Some physicists, including Stephen Hawking and Roger Penrose, actually think that the flow of time, as well as the arrow of time, could be the result of fundamental temporal cosmic irreversibility—that is to say, the expansion of the universe. They would thus correspond with "emerging" properties of the theory. In some simplified versions of quantum cosmology, the volume of the universe is used as a cosmic clock. A question follows: if the universe, after its expansion, entered a contraction phase, would this imply a reversal in the flow of time? (See Stephen Hawking and

Roger Penrose, *The Nature of Space and Time* [Princeton, NJ: Princeton University Press, 2000].)

7 It has already been explained that the concept of the speed of time does not have much meaning, since any speed is a by-product in relation to time. The concept of "the acceleration of time" is not clearer, since any acceleration is a by-product of a speed in comparison to time.

chapter 18

1 We could think that by defining time as "the number of movements according to the before and the after," Aristotle proposed a discontinuous conception of time. This would be forgetting that he immediately added that he thought time was "continuous" because it belonged to a continuum, which is movement, conceived by Aristotle as an unachieved act in the process of becoming. So that the idea of "number" cannot make us think about a "discrete" quantity like the whole numbers 1, 2, 3, and so forth, but what we today call a "magnitude," a term Aristotle reserved for spatial magnitude.

2 Invariance by rotation in space is at the origin (in accordance with the Noether theorem presented earlier) of the conservation of what we call the "kinetic moment"—for example, what makes an ice skater spin quicker when she brings her arms closer to her body.

chapter 19

1 Some symmetries can apply at any point in space-time outside of physical laws. Let's take the example of electromagnetic interaction. To describe it we use the notion of

potential, which is a function defined at every point in space by a number. From the data of this potential we deduce the value of the electromagnetic field at any point. But in the same way an infinite family of parallel straight lines have the same slope, there exists an infinite of potentials, which give the same field. All these potentials are identical *(à une "origine près")*; if we transform one potential into another from the same family, the equations are not affected. The consequence is that if we plugged the cosmos into an electric terminal with a constant electromagnetic potential, every physical phenomenon would be unmodified, hence the term "gauge invariance" to indicate this property.

2 An excellent presentation of the standard model of particle physics is given in Roland Omnès, *Alors, l'un devint deux. La question du réalisme en physique et en philosophie des mathématiques* (Paris: Flammarion, 2002), 181–244.

3 Several superstring theories have been developed over the past thirty years. This multiplicity of theories has created several problems of coherence: the idea of superstring theory was proposed to build a unifying theoretical frame that would encompass all others. But in 1994 some order returned to the house. It was demonstrated that each of the proposed versions was a particular case of a more general theory, called "*M* theory," which remains to be elaborated. The unification of the five superstring theories has been possible thanks to the existence of symmetries, known as "dualities," that link these different theories to one another. More precisely, a certain type of duality links certain theories two-by-two whose "coupling constants" (parameters that measure the interactions between superstrings) are the inverse of each other.

4 For the closed strings, there is always a propagation mode

that corresponds to the graviton, which is the quantum version of Einstein's gravitational field.

5 Theodor Kaluza, "Zum Unitätsproblem der Physik" (On the problem of unity in physics), *Sitzungsberichte der Preussische Akademie der Wissenschaft* (1921), 966–72; Oscar Klein, "Quantum theory and 5-dimensional theory of relativity," *Z. Phys.* 37 (1926), 895.

6 In order to be completely fair, we should say that the physicist Georges Gamow had already considered the existence of a fifth dimension. In his first article, published in 1926, he proposed to interpret the wave function of the Schrödinger equation as the equivalent of a fifth dimension, which should be added to the four usual dimensions of space and time. This suggestion, as we know it, did not convince many people.

7 Actually, another possibility says that the additional dimensions are infinite, but that we cannot cross through them. The *branes* (a word derived from "membrane" that enables us to speak about *n*-branes for *n* dimensions subspaces) are subvarieties of the fundamental space on which are tied the open strings. In this frame, our universe would be a three-dimensional flag floating in this bigger fundamental space where the closed strings are living (like the graviton), a flag on which the opened strings—that is to say, the particles of the standard model—would be condemned to live.

8 After the Planck length, the quantic fluctuation of space-time becomes so important that the notions of distance, mass, and energy cannot keep their usual meaning.

9 One teraelectronvolt (TeV) is worth 10^{12} electronvolts, or $1.6\ 10^{-7}$ joules.

10 See Ignatios Antoniadis, "Et si l'on prouvait la théorie des superstrings?" in *La Recherche,* hors série no. 8, July/August/September 2002.

11 It has been demonstrated that if there exists at least one additional dimension traveled by light at the scale of 10^{-18} meters, we should observe some particles, known as "Kaluza-Klein states," at the LHC similar to photons but with a mass that is higher than the size of the additional dimensions would be smaller. We will actually detect pairs of electron-positrons, or muon-antimuons, coming from the disintegration of these particles.

12 For more on this point, see J. Richard Gott, *Time Travel in Einstein's Universe: The Physical Possibilities of Travel Through Time* (Boston: Houghton Mifflin, 2001).

chapter 20

1 There is one word we cannot include in this list—mass. Indeed, the standard model of particle physics proposes a scenario that explains how particles, originally without mass, managed to acquire mass as a result of a "mechanism of the spontaneous breaking of symmetry," also called the "Higgs mechanism." This hypothesis will soon be tested at the LHC of the CERN.

2 Emmanuel Kant, *Prolégomènes à toute métaphysique future qui voudra se presenter comme science* (Paris: Vrin, 1968), 132.

3 A simple argument provides the scales of time and distance within which a conceptual revision should imperatively intervene, enabling us to imagine quantum physics and general relativity together. This argument is noted because some fundamental constants exist in physics: gravitation, *G;* the speed of light, *c;* and the Planck constant, *h.* With each of these three constants expressed according to a well-

defined unity, it is possible to combine them in order to get a magnitude expressed in a unit of time. The length obtained this way, called the Planck length, equals $(Gh/c5)^{1/2}$. It is approximately worth 10^{-43} seconds. Beyond this scale, our usual representations of space and time lose any signification and known alternatives remain highly speculative.

4 To know more about these questions, see Hawking and Penrose, *Nature of Space and Time.*

5 This question is reminiscent of the one asked ironically by the neo-Platonists to Saint Augustine, who defended the idea of a temporal beginning of the universe: "What was God doing before the creation of the universe?" Saint Augustine would reply that there was no time before the creation of the universe, since time is only a property of the universe, and God, in his eternity, is exempt from it. In other words, there cannot be any empty time, time without a world to unfold in.

chapter 21

1 Minutes of the meeting of the Société française de philosophie, held April 6, 1922 (*La Pensée,* n. 210, February–March 1980, 22).

2 Paul Valéry, *Oeuvres* (Paris: Gallimard, coll. "Bibliothèque de la Pléiade," t. II, 1984), 713.

3 On this point Marcel Proust said: "The days preceding my dinner with Mme de Stermania were not delicious, but unbearable. In general, the shorter the time separating us from what we propose to ourselves, the longer it appears to us, because we look at it in shorter measures or simply because we think of measuring it." (Marcel Proust, *Le côté*

de Guermantes [Paris: Gallimard, coll. "Bibliothèque de la Pléiade," vol. 2, 1954], 172.)

4 Fraisse, Paul. *Psychologie du temps.* "Bibliothèque scientifique internationale psychologie." Paris: PUF (1957).

5 See Viviane Pouthas, "Où sont les zones du temps dans le cerveau?" in *La Recherche*, hors série n. 5, April 2001, 80–83.

6 We know, for example, that short durations tend to be overestimated, long ones underestimated, and that intense stimulation always seems to last longer than a less intense stimulation with the same duration. Our estimation of durations is also subject to a distortion that depends on the nature and the modulation of the signals that we receive. For example, a gap between two short sounds is always thought to be shorter than a gap of time with the same duration but "filled" with a continuous sound. Sounds generally seem to last longer than light signals of the same duration. For our brain, sounds seem to drag things out more than glimmers or flashes, as if they were wrapping themselves in an increased temporal remnant, whose cause remains poorly understood.

7 Gaston Bachelard, 33.

8 Several thinkers have tried in vain to find the unifying point from which the *physis* and the *psyche* could arise by differentiation. This was also the case of Wolfgang Pauli, one of the founding fathers of quantum physics, who had a long correspondence with the psychiatrist and psychoanalyst Carl Gustav Jung. Reading their letters, we see that both were convinced that it was as impossible for the psychologist to neglect the methodical principles of physics as it was for the physicist not to take into account his psychological experiences. They then agreed that the single acceptable approach was one that recognized the compatibility of

the two sides of reality, the physical and the psychic. All that remains is to demonstrate how these two sides can be reconciled. (See Wolfgang Pauli and Carl Gustav Jung, *Correspondance, 1932–1958* [Paris: Albin Michel, coll. "Sciences d'aujourd'hui," 2000].)

chapter 22

1 Maurice Merleau-Ponty, *Phénoménologie de la perception* (Paris: Gallimard, 1995), 474.
2 Saint Augustine, *Confessions, trad. L. de Mandalon* (Paris: Seuil, 1962), XI, 15.
3 In a more general way, we could say that the specific trait of consciousness is never to be totally present to the present, as Blaise Pascal had noticed: "We are never satisfied with the present time. We anticipate the future as being too slow to arrive, as if to hasten its flow; or we summon the past to stop it as if it were too quick . . . Let everyone examine his thoughts, he will find all thoughts busy with the past and the future. We hardly think about the present; and if we do, it is only to take its light to prepare the future. The present is never our objective: the past and the present are our means; only the future is our goal. That's how we never live, but we hope to live; and, always preparing to live, it is inevitable that we will never live." (Blaise Pascal, *Pensées,* éd. Brunschvicg, frag. 172; éd. Lafuma, frag. 47)
4 Thomas d'Aquin, *Summa contra Gentiles,* Lib. I, cap. LXVI. This conception of Saint Thomas Aquinas will be taken up by another academic of the time, Pierre Auriol, who wrote: "Some are using the image of the centre of the circle in its relationship to all points on the circumference; and they

affirm that it is the same as the *nunc* of eternity in its relationship with all parts of time. Eternity, they say, currently coexists with all of time." (*commentarii in Primum Librum Sententiarum Pars Prima,* [Rome, 1596], 829.)

chapter 23

1 Sigmund Freud, *L'Interprétation des rêves* (Paris: PUF, 1967), 526.
2 See Sylvie Le Poulichet, *L'oeuvre du temps en psychanalyse* (Paris: Rivages, 1994).
3 André Green, *Le Temps éclaté* (Paris: Editions de Minuit, 2000).
4 Sigmund Freud, *Essai sur l'unconscious*, in *Oeuvres completes,* XIII (Paris: PUF, 1988), 226.

chapter 24

1 Clément Rosset, *Le Monde et ses remèdes* (Paris: PUF 2000), 143.
2 André Breton, *L'Amour fou* (Paris: Gallimard, coll. "Folio," 1991), 171.
3 Thus Proust's famous "temps retrouvé" can be interpreted as a deliverance toward the mobile. The joy of Proustian reminiscence is like a platform from which we can eventually grasp past moments that have escaped us. By linking dream and memory it offers the possibility of an imaginary stop, and it does not matter if this "temporal balcony" is as illusory as the fusion with an "other."
4 "I realized all that the imagination can hide behind a small

piece of a face," says the narrator of *La Recherche du temps perdu* (Proust, *Le côté de Guermantes,* 159.)

chapter 25

1 Spinoza, *Ethique*, II, prop. 44, corollaire 2 (Paris: Gallimard, coll. "Bibliothèque de la Pléiade"), 638.
2 G. Galilée, *Le Messager des étoiles* (Paris: Seuil, 1991), 151.
3 G. Galilée, *Dialogue et lettres choisies* (Paris: P. H. Michel and G. De Stantillana, 1966), 37.
4 This is all the more so since as an experience, death is not livable; it does not constitute a present for the one dying, for the one "living" it. We remember the old truism from Epicurus: "When you are here, death is not here; when it is here, you are not here." Death happens in time as a destructive singularity of the being outside of the duration of the being.
5 Georges Canguilhem, *Idéologie et rationalité dans l'histoire des sciences de la vie* (Paris: Vrin, 1981), 138.
6 More than two millennia ago, when atoms were still metaphysical entities, Lucretius already assumed that we could laugh without being formed with laughing atoms and that we could philosophize without being formed with philosopher atoms (Lucrèce, *De rerum natura,* livre II, 985–90).
7 Kant writes: "The inertia of matter is and does not mean anything else than the *absence of life* of matter in itself. Life is the power of a substance to determine how to act according to an *internal principle* . . . Yet, we do not know any internal principle in a substance other than *desire,* and in a general way, any other internal activity than *thinking* and what depends on it, the *feeling* of pleasure and pain, the appetite or the will.

Though these principles of determination and these actions are not part of the representations of the external senses nor consequently the determination of matter as such. And, any matter as such is deprived of life." (Emmanuel Kant, *Fondements métaphysiques de la science et de la matière*, trad. J.Gibelin [Paris: 1900], III, 130–31).

8 A radioactive atom is an atom that has gathered too much energy as it forms. At one moment or another, it has to get rid of this surplus by releasing particles and by transforming itself into another atom at the same time.

9 A few figures are enough to demonstrate that we do not die according to the same temporal law as radioactive atoms: if the mortality rate was constant for humans, with, for example, a half-life of 75 years (the life expectancy in developed countries), one-fourth of each age class would reach 150 years old, and one person out of a thousand would reach 750 years old. We see that the gap with reality is huge. It is explained by the fact that the human mortality rate increases with age. Note, however, that there is no parity in this field: in France there are seven to eight times more female than male centenarians, which seems to give reason to Pierre Dac when he states that "women live longer than men, especially the widows."

10 The finiteness of the famous *Dasein*—"the being of this being that we know as human life"—would be the very foundation of its existence and not an accident of its immortal essence. In sum, the human being gives himself his time because he goes ahead of his death; he is continually expecting death.

11 This is a reference to the expression *ça sent le sapin,* or "he's got one foot in the grave."

12 "To avoid feeling the horrible burden of Time which breaks

your shoulders and bends you towards the earth, you have to unceasingly get drunk. But with what? Wine, poetry, virtue, as you wish. But get drunk!" (Charles Baudelaire, "Get drunk," poem XXXIII, *Le spleen de Paris* [Paris: Flammarion, coll. "GF-Flammarion," 1987].)

13 Of course the problem is that this "almost infinitely" actually means "not infinite at all."